A CALIFORNIAN'S GUIDE TO THE TREES AMONG US

A CALIFORNIAN'S GUIDE TO THE TREES AMONG US

Second Edition
Expanded and Updated

Matt Ritter

Heyday, Berkeley, California

This project was made possible in part by a generous grant
from the Stanley Smith Horticultural Trust.

Library of Congress Cataloging-in-Publication Data

Names: Ritter, Matt, author.
Title: A Californian's guide to the trees among us / Matt Ritter.
Description: Expanded and updated. | Berkeley, California: Heyday, [2022] |
    Includes bibliographical references and index.
Identifiers: LCCN 2021970004 (print) | LCCN 2021058707 (ebook) | ISBN
    9781597145602 (paperback) | ISBN 9781597145701 (epub)
Subjects: LCSH: Trees--California--Identification. | Guidebooks.
Classification: LCC QK149 .R58 2022 (print) | LCC QK149 (ebook) | DDC
    582.1609794--dc23/eng/20220114
LC record available at https://lccn.loc.gov/2021970004
LC ebook record available at https://lccn.loc.gov/2021970094

Cover design: Ashley Ingram
Cover photography: Matt Ritter
Illustrations: Annette Filice and Nayl Gonzalez
Interior Design and Typesetting: Leigh McLellan Design and Matt Ritter,
    with Ashley Ingram and Marlon Rigel

Published by Heyday
P.O. Box 9145, Berkeley, California 94709
(510) 549-3564
heydaybooks.com

Printed in East Peoria, Illinois, by Versa Press, Inc.

10 9 8 7 6 5 4 3 2 1

This book is dedicated to my inimitable and rowdy son,
Abel Ahimsa Ritter, on his fifteenth birthday.
Where did you come from?

Contents

Acknowledgments for the Second Edition   *ix*

Foreword   *xi*

Introduction to the Second Edition   *1*

Identification Key to Trees Commonly Cultivated in California   *12*

Compendium of Trees
*(Within the following three categories,
trees are arranged alphabetically by genus)*

       Gymnosperms   *27*

       Angiosperms   *45*

       Palms   *163*

Appendix A: Changing Plant Names   *173*

Appendix B: Locations of Photographed Trees   *175*

Glossary   *177*

References, Further Reading, and Great Tree Books   *183*

Index   *185*

About the Author   *195*

Tree Identification Flowchart   *196*

Acknowledgments for the Second Edition

I have attempted to make this book as botanically accurate, rigorous, and comprehensive as possible. Many people have given me advice and guidance during its preparation (and my career in general). Any errors in the book are entirely my own. Thank you to Annette Filice for creating all the drawings in this book, and to Eric Meyer and Enrica Lovaglio Costello for help and feedback on the book's design.

I've had the honor of interacting with some of California's most expert and prolific plantsmen and plantswomen. Each has taught me an incredible amount about trees in California and trees in general. These friends and colleagues include Randy Baldwin at San Marcos Growers; Ken Greby, who, in my opinion, has more general knowledge about trees than any other human; palm experts Jason Dewees and Don Hodel; Brett Hall and Melinda Kralj at the UC Santa Cruz Arboretum; Warren Roberts at the UC Davis Arboretum; Brian Kemble at the Ruth Bancroft Garden in Walnut Creek; Richard Borevitz in San Diego; Dylan Hannon and Kathy Musial at the Huntington Botanical Gardens in San Marino; Arthur Lee Jacobson, author of *Trees of Seattle*; arborist extraordinaire Kevin Eckert; and Mike Sullivan, author of *The Trees of San Francisco*. Thank you to Rose Epperson and the membership of the Western Chapter of the International Society of Arboriculture for their continued interest and support.

Several exceptional students and friends helped with the book, including Kieran Althaus, Cami Pawlak, Jason Johns, and Paul Excoffier. Thank you to my colleagues Dr. Natalie Love and Sean Ryan for their continued guidance, feedback, and help on this and many other projects. Thanks go to Heyday's editorial and design staff, particularly Marlon Rigel and Marthine Satris.

It has been a great honor to know and interact with Peter Raven. His life and work have inspired me and an entire generation of plant biologists, and I am grateful for his encouragement and endorsement of this project and my career as a botanist.

Several people spent many hours reading and correcting versions of the manuscript for both the first and second editions. I greatly appreciate the efforts of Sairus Patel for so carefully scrutinizing the writing and finding even the most difficult-to-find errors. I extend my deepest gratitude to my botanical mentor Dr. David Keil, professor of botany at Cal Poly, San Luis Obispo, from whom I've learned so much. He improved the book significantly with effort, time, ideas, and encouragement.

I am grateful for having the close collaboration, friendship, and enthusiasm of my colleague Dr. Jenn Yost. It was through discussions with her that the idea for this book arose. Above all, and as always, my most profound thanks go to my patient and incomparable wife, Sarah Allen Ritter, without whom I would be capable of next to nothing.

Many people, other than the authors, contribute to the making of a book, from the first person who had the bright idea of alphabetic writing through the inventor of movable type to the lumberjacks who felled the trees that were pulped for its printing. It is not customary to acknowledge the trees themselves, though their commitment is total.

—Richard Forsyth and R. Rada

Foreword

Peter Raven
President Emeritus, Missouri Botanical Garden

It used to be said that you could grow anything in California if you watered it, and that is true to a rather incredible degree. The conditions that greeted Fray Junipero Serra and his men when they entered what is now California in the summer of 1769 were similar to those they knew in their native Mallorca: evergreen drought-resistant trees and shrubs, grassy slopes that turned brown in the summer, and that wonderful potential for growing almost anything if water was available.

And grow they did. In California's gardens and parks and along its streets there exists today one of the most complete and interesting sets of trees to be found anywhere in the world. The largest proportion come from Australia, where the summer-dry climate in many areas is similar to that of California. Many come from temperate East Asia, the richest part of the Northern Hemisphere biologically; others come from temperate to subtropical regions all over the world. Needless to say, there are many more kinds of trees that could be cultivated in California, with care to avoid those likely to become weedy and invasive, thereby endangering native plants and animals. All in all, the diversity of cultivated trees in California creates a perfect setting for the wondrously eclectic life of the Golden State.

This attractive and well-produced book presents a virtual arboretum of the tree species— both familiar and extraordinary—that make up the full set of California's cultivated trees. Stately or striking, with lovely flowers, spectacular foliage, or other endearing features, they bring beauty and grace to our lives each month of the year, in all of the state's myriad landscapes. Just as the author of this book grew up in the valley oak landscape of Northern California, all Californians grow up in settings dominated by the trees characteristic to their particular area—and if they are fortunate, they are enriched by their knowledge of these trees. For every Californian alive in the mid 1930s, when I was born, there are now nearly eight times as many. No wonder California seems more crowded than it did then, a trend that has accelerated the desire to plant beautiful trees: they create space and context for our lives, and peace within us.

Turning the pages of this book, I realize how valuable it would have been to me as a child, growing up in San Francisco and gradually coming to realize the dimensions of the natural diversity that surrounded me. When I visited Southern California, I became acquainted with a whole new set of cultivated trees and other plants, and I began to understand the differences between the foggy north and the sunnier south. For those enjoying similar experiences now, *A Californian's Guide to the Trees among Us* will be a useful and inspiring companion. Trees invite us—and our children—to learn about nature, creating a source of enjoyment that will last a lifetime; someone who appreciates the diversity of trees will start to discern the seemingly endless variety within nature, and will, perhaps, be a better-informed citizen in all aspects of life.

However, we cannot count on our cultivated trees being with us in perpetuity. Despite recent successes, urban pollution is an ever-present threat to their condition and longevity. Global climate change is threatening California's water supplies, and the thirty-eight million people who now live within the state's boundaries must learn how to conduct their activities in the face of what are likely to be increasing shortages. If the cold waters offshore are eliminated or severely limited by global warming, fog will not nourish the trees of California's coastal areas, exemplified by the lovely coast redwoods, to the degree it does now, and many species of plants and animals will be threatened with extinction. The first six months of 2010 were the warmest six months recorded worldwide since climate measurements were begun, and the trend continues. As the bulk of trees is reduced worldwide, more carbon dioxide, the principal man-made greenhouse gas, reaches the atmosphere and the climate changes still more.

Trees are often harvested for their wood or another of their many purposes, or simply cleared to be replaced with houses or agriculture, becoming all too threatened in their native landscapes. Some are very rare now, and others are becoming rare as their habitats are destroyed. To a degree, the set of trees that adorns the California landscape constitutes a kind of Noah's ark for the survival of those species.

Among the 150 kinds of trees described in this handsome book are all of the most widespread and familiar species that are cultivated in different parts of the state. Matt Ritter is intimately familiar with these trees. The degree of his knowledge shows through on every well-illustrated page, and his appreciation is evident in the detailed, informative accounts of individual species. A popular and accessible treatment, the book will be useful for all citizens. Arborists, nurserymen, and consultants will find it helpful for identifying trees and for insight into what trees might be suitable for future plantings. In fact, any reader can use the identification keys and clear descriptions to identify trees and learn about their unique characteristics.

Inspiring this generation and the next with a love of the natural world, books like *A Californian's Guide to the Trees among Us* encourage the re-greening of California in ways that are sensitive to its fundamentally limited water supply. They feed the groundswell for conservation that has grown steadily since the time of John Muir and is now a persistent feature of the vision that Californians have for the future. In that sense, Matt Ritter has done us a wonderful favor, and we will thank him for his industry and for sharing his abundant visual and verbal skills.

Introduction to the Second Edition

I have always been drawn to trees and believe that I am not alone in this sentiment. Growing up among the remnant stands of large valley oaks in a small interior valley of California's North Coast Range, before I ever considered myself a botanist, I knew there was something inspiring about trees. A magical quality surrounds a large or exotic tree. As I grew older and my interest in plants became more refined, I never escaped the spell cast on me by trees.

Those who live in California need not travel to exotic places to see an eclectic mix of trees from all corners of the earth; one needs only to stroll down a local street and look up. California's agreeable climate and rich horticultural history have converged to populate our towns and cities with trees that reward those who notice them with vibrant color, bizarre shapes, unusual textures, and unexpected smells. During my early adulthood, I dedicated many trips in California and beyond to finding impressive or strange trees—often to the dismay of less interested travel companions. It was on these trips that I came to appreciate the astounding diversity of trees in our state. Besides an abundant array of native trees, which number about one hundred for the state, California has park, garden, yard, and street trees—our so-called urban forest—that are among the most splendid and varied in the world. These beautiful organisms are all around us, vital to our well-being and worthy of our praise and fascination.

Red Oak acorn (*Quercus rubra*)

Urban landscapes are not easy places for trees, yet trees are essential to the quality of life of the humans around them. Trees are giving and forgiving. As they persevere, sometimes in the most adverse of conditions, they beautify our world. They muffle noise, cool neighborhoods, create wildlife habitat, mitigate pollution and runoff, conserve energy, and make urban living healthier and more peaceful. It is no wonder that city dwellers place such great value on their trees. They must survive in the face of root constriction and compaction, air pollution, interference from overhead wires, neglect, random acts of arboreal violence, and many months without rain. The physical beauty of many of California's most charming cities is due, in no small part, to the well-tended trees lining their streets.

Tree-lined sidewalks in (left to right) Davis, Los Angeles, Palo Alto, and Santa Barbara

Trees on Yerba Buena Island frame the San Francisco–Oakland Bay Bridge

California's climate creates a wide range of growing conditions for trees. The dry, practically frost-free areas of coastal San Diego; the warm and sunny Los Angeles Basin; the opulent and balmy gardens of Santa Barbara; the seasonally distinct, rich plains of the Central Valley; and the foggy, windswept, sandy hills of San Francisco all favor certain types of trees.

With its reverence for the outdoors and desire for greener cities, California's populace has taken advantage of these diverse growing conditions for over 150 years. Since California's first nurseries were founded, the state's residents have sought out novel and unknown trees from foreign lands. Many species were tried, many failed, but a significant number succeeded and now lavishly inhabit our streets, parks, and gardens. There are risks associated with the sustained introduction of exotic trees, and some trees have succeeded in invading California's wildlands. Be that as it may, California's cities have become conservatories for the world's tree biodiversity.

There are approximately 350,000 species of plants globally, and about 60,000 of these grow as trees. California is home to thousands of different species of ornamentally grown trees, most of these as isolated specimens in our many arboreta and botanical gardens. However, in many of California's municipalities, the same array of 150 or so species is widely planted, with exceptions primarily due to the frost tenderness of species successfully grown only in coastal or Southern California.

This book is a natural history and identification guide to these species: the most commonly grown trees in urban and suburban landscapes in California. It is no accident that these trees are the most frequently encountered in California; they are here because they have earned the favor of horticulturists, city planners, or the general public. They may be beautiful, but they are also resilient and easy to grow.

Southern Magnolia
(*Magnolia grandiflora*)

It is not always because of their desirable practical traits that these trees are among the commonly cultivated. There are other reasons, often complicated, involving tradition, nostalgia for places from which many Californians have emigrated, historically common trade routes, similar climates, and, occasionally, the random and idiosyncratic interests of tree connoisseurs.

Each of these commonly grown trees has a history and a story. I have attempted to chronicle parts of those stories in these pages. Many other trees are discussed briefly and included in the

The palm tree collection at the Santa Barbara County Courthouse Sunken Gardens

key to species, but admittedly, this is an incomplete guide to the immense variety of trees in California.

As this book focuses on the trees of urban and suburban environments, I have included California native trees, such as coast redwood (*Sequoia sempervirens*) and Monterey cypress (*Hesperocyparis macrocarpa*), only if they are widely cultivated. I purposely omitted most shrubs, even those painstakingly pruned into single-trunked trees, as well as trees that are primarily grown agriculturally or in home gardens for their fruits or nuts.

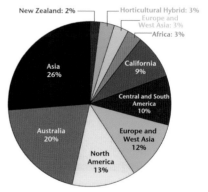

Origins of California's Urban Tree Species

The approximately 150 commonly grown species are found in about ninety genera and thirty-five families. The myrtle, legume, and rose families are the most well represented, but the species are widely distributed among plants. California's commonly cultivated trees have a wide variety of geographic origins, the most common being Australia.

Trees embody the strength and beauty of the natural world. The rich and subtle connections that link humans and trees have undoubtedly existed since our primitive beginnings in arboreal habitats and the origin of human consciousness. Countless writers and thinkers have tried to interpret the obscure wisdom and ancestral nostalgia that trees seem to offer us. Trees provide a central theme in literature, art, pop culture, mythology, and religion.

Evolutionary Relationships among California's Commonly Cultivated Trees

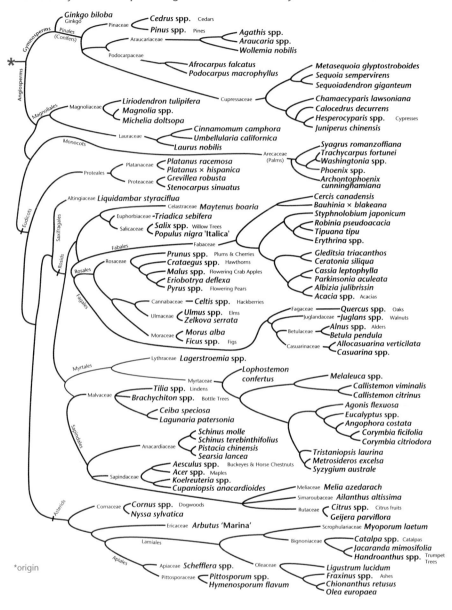

*origin

In many cultures, trees symbolize community, fertility, life, and the interdependence of the natural world. In cities, trees are often poignantly contrasted with stark buildings, sidewalks, streets, and gutters, silently reminding us of lost connections with nature and our longing for a time before our taming and civilization. Most pages in this book include a quote, an idea, or a famous piece of writing about trees. Some of these are meditations on the meaning of life that trees seem to hold; some recount the story of the deep connection between humans and the trees around us; others are just funny or thought-provoking.

Moreton Bay Fig
(*Ficus macrophylla*)

Tulip Tree
(*Liriodendron tulipifera*)

Italian Stone Pine (*Pinus pinea*)

I have attempted to combine science, natural history, drawing, photography, and prose in this book with the hope that you will gain greater appreciation and reverence for the trees around us. Since 2011, when the first edition of this book was published, much has changed in our world. The protection that trees provide us is more valuable than ever. Trees are a gift, many planted by predecessors long past, that can be honored only by valuing them, protecting them, and growing new ones for future generations.

This is a book about trees, made from the bodies of dead trees, and reading it is a poor substitute for experiencing these incredible organisms directly and personally. Take it with you and walk out among the trees in your neighborhood. Use it to identify unknown trees, watch them change throughout the year, place a hand on the bark, feel the leaves, and smell the flowers. Pay close attention to them, fraternize with them, commune with them, and, most important, appreciate them. The time spent learning, watching, and experiencing these wondrous organisms will be well worth the effort.

Learning to Look at Trees

There is a joy that comes from identifying and learning more about the organisms around us—a feeling I have with each new tree species I meet. The barriers to experiencing this joy can be overcome by learning how to look at trees and what features are important for their identification. Once your eyes and mind become sharpened with the ability to distinguish important differences and similarities in trees, these organisms cease to be simply a green background, and a new world comes to life.

As the beauty in the details of trees is unveiled, learning new species becomes easier and easier. Botanists love words, sometimes seemingly more than we love plants, occasionally using technical words to describe relatively simple concepts. Throughout this book, I have minimized the use of botanical jargon. However, there are botanical terms that eloquently describe parts of trees, and these words and the structures they describe are a key to entering the world of trees. Technical terms are defined in detail in the glossary. What follows here is a brief discussion of the bark, leaf, fruit, cone, and flower characteristics that are referenced in the discussions of the species. Warning: you may be asked to speak to a tree.

Japanese Flowering Cherry blossom (*Prunus serrulata* 'Kwanzan')

Mayten tree (*Maytenus boaria*)

Clockwise from left: bark from *Melaleuca ericifolia*, *Arbutus* 'Marina', *Platanus × hispanica*, *Betula nigra*

A tree can be defined as a woody plant, usually with a single main stem (the trunk) and a network of branches spreading from that trunk to form a distinct and elevated crown. What is the difference between a tree and shrub? A tree will hurt you if it falls on you.

The living tissues within the trunk are protected by the outer layers of tissue of the bark. The appearance of the bark varies greatly in different species and can often be one of the most conspicuous and distinctive features of a tree. As an example, look at the difference between the white iron bark (*Eucalyptus leucoxylon*, page 86) and the red iron bark (*Eucalyptus sideroxylon*, page 86).

Trees in this book are described as being either deciduous, evergreen, or partly deciduous. *Deciduous* trees lose their leaves all at once, at some point during their annual cycle. Most often, this is during the cool, wet California winter, although there are some trees, such as jacarandas (*Jacaranda mimosifolia*), that lose their leaves during the spring and early summer. *Partly deciduous* trees, such as the floss silk (*Ceiba speciosa*), may lose leaves from only a portion of the crown, but may, on occasion, lose all their leaves. *Evergreen* trees, such as eucalypts, retain their leaves year-round, with leaves lasting from one to several years and never falling all at once.

The most varied, conspicuous, and characteristic organs of trees are their leaves. Leaves arise from the stem at regular points called *nodes* and are generally composed of a leaf stalk (the petiole) and a flat, expanded area called the blade. A bud (usually called an axillary bud) is always present and usually evident just above where a leaf attaches to the stem at the node.

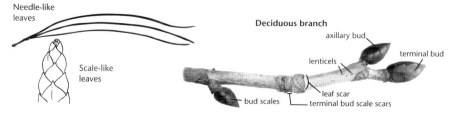

Needle-like leaves

Scale-like leaves

Deciduous branch

axillary bud

terminal bud

lenticels

leaf scar

bud scales

terminal bud scale scars

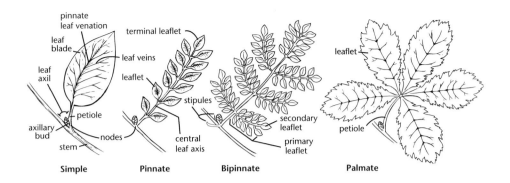

pinnate
leaf venation
terminal leaflet
leaf
blade
leaf veins
leaf
axil
leaflet
stipules
axillary
bud
petiole
nodes
stem
central
leaf axis
secondary
leaflet
primary
leaflet
leaflet
petiole

Simple   **Pinnate**   **Bipinnate**   **Palmate**

As you begin a new relationship with an unknown tree, you should ask it two essential questions, the answers to which will do wonders for helping you learn more about it. The first question is "What is your leaf form?" There are three main options for leaf form. Most trees have *simple* leaves, with undivided blades. The blade may be deeply lobed, as in maples and oaks, but is still considered simple if it is not divided into leaflets. Other trees have compound leaves, which are divided into leaflets in two ways. *Palmate* leaves have leaflets spreading from the tip of the leaf stalk, as on horse chestnuts and buckeyes (*Aesculus* spp.); *pinnate* leaves have leaflets arranged along the edges of a central axis in a feather-like fashion, as on ashes (*Fraxinus* spp.). The leaflets are divided again in *bipinnate* leaves, as on jacarandas (*Jacaranda mimosifolia*). It can be difficult to distinguish between a leaf and a leaflet, but it helps to know that a leaflet never has an axillary bud.

The second question for your unknown tree is "How many of your leaves are attached to each node?" Regardless of their form, leaves are attached to the stem in a manner that is *alternate* (with one leaf at each node), *opposite* (with two leaves attached to each node across the stem from each other), or *whorled* (with three or more leaves attached around the stem at each node.) Once you have answered these two leaf questions, you are well on the way to knowing your tree. Still, a more intimate look may be required, in which case you must explore how the tree reproduces.

With only a very few exceptions (such as the ginkgo), trees can be divided into two groups that reproduce in different ways: the *conifers* (cone-bearing trees, sometimes called softwood trees) and the *angiosperms* (trees that make flowers, sometimes called hardwood

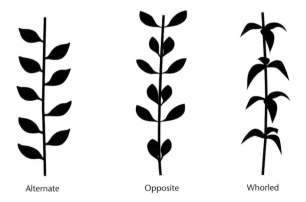

Alternate   Opposite   Whorled

Deodar Cedar cone
(*Cedrus deodara*)

Cook Pine cone
(*Araucaria columnaris*)

trees). Conifers bear their reproductive structures in male (pollen-forming) and female (seed-forming) cones; their cones are never hermaphroditic, as is the case with many flowers. California is famous for its record-holding conifers, with the world's tallest (coast redwood, *Sequoia sempervirens*), oldest (bristlecone pine, *Pinus longaeva*), and most massive (giant sequoia, *Sequoiadendron giganteum*) trees all living natively within the state. In California's urban landscapes, conifers are cherished for their stately appearance, often lending a look of grandeur and antiquity to the parks and gardens where they are planted. Species from five of the eight conifer families are widely cultivated in the state's metropolitan areas, and about fifteen species of conifers are described here in detail. Although conifers were Earth's dominant plants for millions of years, many have become relics and are now restricted to only a fraction of their once expansive ranges. Our official state tree (the coast redwood), for instance, was at one time found throughout the Northern Hemisphere and is now limited to the mild habitats of California's northern coast. In most areas, as conifer dominance waned, flowering plants flourished.

Trees that make flowers and fruits, the angiosperms, are the second and vastly more numerous group. The flowers and fruits found on trees are not primarily for our enjoyment and sustenance; they are sexual organs that have been naturally selected, and they are marvelously efficient at facilitating pollen movement (usually by wind or animals), fertilization, and the production and dispersal of seeds. Flowers found on California's trees vary greatly, ranging in color and size from cryptic and barely visible on the sheoaks (*Casuarina* spp.) to showy and larger than a dinner plate on southern magnolia (*Magnolia grandiflora*).

A typical flower is composed of four concentric rings, or whorls, of floral organs. Just inside the outermost ring of *sepals* (collectively called the *calyx*), which are usually green, is the often showy and colorful whorl of petals (collectively called the *corolla*). Further

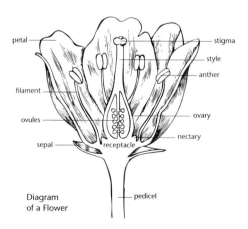

Diagram
of a Flower

Fruit and seeds of Kurrajong (*Brachychiton populneus*)

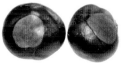

Red Horse
Chestnut seeds
(*Aesculus × carnea*)

Flower Color and Time

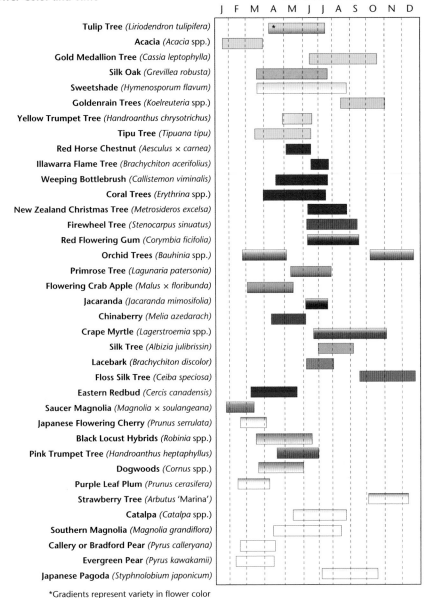

*Gradients represent variety in flower color

inside are the male *stamens*, each with a pollen-filled *anther* at its tip, and in the center are the female *carpels*. Each carpel has an *ovary* that completely encloses the *ovules* (unfertilized seeds) and on top of which sits a projecting *style* with a pollen-receptive *stigma* at its tip. Some tree flowers lack one or more of these rings of parts. Many wind-pollinated tree flowers lack petals.

Regardless of the many modifications and adaptations that flowers have undergone, knowing the four whorls will help you identify an unknown structure based on its position in relation to other identifiable whorls. Only rarely do flowers occur individually on trees; usually they are borne in specifically arranged clusters called *inflorescences*. The majority of trees make bisexual flowers, with functioning male and female organs in each flower. Species that make unisexual reproductive structures (cones or flowers) may have male and female parts on the same tree (*monoecious*, pronounced moan-EE-shuss), or only one sex per individual tree (*dioecious*, pronounced die-EE-shuss).

Tree identification can sometimes be as easy as noting the floral color and time of flowering, but it may be more difficult and require close observation of many aspects of the tree. In these cases, refer to the Identification Key to Trees Commonly Cultivated in California.

How This Book Works

The species in this book are listed in alphabetical order by genus in three large groups: conifers; broad-leaved or flowering trees; and palms. To identify an unknown tree, there are two paths to take. If you know beforehand that the tree belongs to a specific group within those categories, such as the conifers, maples, oaks, or palms, you can turn to that section of the book and look through photos to find a match. If the tree is completely unknown to you, start with the identification key on page 12, which uses readily observable characteristics to help identify the tree.

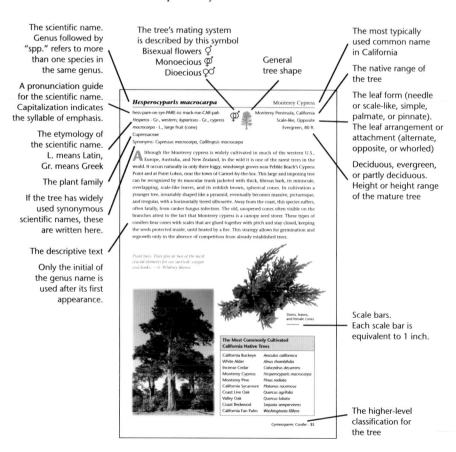

The scientific name. Genus followed by "spp." refers to more than one species in the same genus.

A pronunciation guide for the scientific name. Capitalization indicates the syllable of emphasis.

The etymology of the scientific name. L. means Latin, Gr. means Greek

The plant family

If the tree has widely used synonymous scientific names, these are written here.

The descriptive text

Only the initial of the genus name is used after its first appearance.

The tree's mating system is described by this symbol
Bisexual flowers ♀
Monoecious ♂
Dioecious ♀♂

General tree shape

The most typically used common name in California

The native range of the tree

The leaf form (needle or scale-like, simple, palmate, or pinnate). The leaf arrangement or attachment (alternate, opposite, or whorled)

Deciduous, evergreen, or partly deciduous. Height or height range of the mature tree

Scale bars. Each scale bar is equivalent to 1 inch.

The higher-level classification for the tree

Hesperocyparis macrocarpa — Monterey Cypress

hess-pare-oe-sye-PARE-iss mack-roe-CAR-pah — Monterey Peninsula, California
Hesperos - Gr., western; *kyparissos* - Gr., cypress — Scale-like, Opposite
macrocarpa - L., large fruit (cone) — Evergreen, 80 ft.
Cupressaceae
Synonyms: *Cupressus macrocarpa, Callitropsis macrocarpa*

Although the Monterey cypress is widely cultivated in much of the western U.S., Europe, Australia, and New Zealand, in the wild it is one of the rarest trees in the world. It occurs naturally in only three foggy, windswept groves near Pebble Beach's Cypress Point and at Point Lobos, near the town of Carmel-by-the-Sea. This large and imposing tree can be recognized by its muscular trunk jacketed with thick, fibrous bark, its miniscule, overlapping, scale-like leaves, and its reddish brown, spherical cones. In cultivation a younger tree, invariably shaped like a pyramid, eventually becomes massive, picturesque, and irregular, with a horizontally tiered silhouette. Away from the coast, this species suffers, often fatally, from canker fungus infection. The old, unopened cones often visible on the branches attest to the fact that Monterey cypress is a canopy seed storer. These types of conifers bear cones with scales that are glued together with pitch and stay closed, keeping the seeds protected inside, until heated by a fire. This strategy allows for germination and regrowth only in the absence of competition from already established trees.

Plant trees. They give us two of the most crucial elements for our survival: oxygen and books. —A. Whitney Brown

Stems, leaves, and female cones

The Most Commonly Cultivated California Native Trees	
California Buckeye	*Aesculus californica*
White Alder	*Alnus rhombifolia*
Incense Cedar	*Calocedrus decurrens*
Monterey Cypress	*Hesperocyparis macrocarpa*
Monterey Pine	*Pinus radiata*
California Sycamore	*Platanus racemosa*
Coast Live Oak	*Quercus agrifolia*
Valley Oak	*Quercus lobata*
Coast Redwood	*Sequoia sempervirens*
California Fan Palm	*Washingtonia filifera*

Gymnosperm: Conifer : 35

Each entry in the book describes a single species or a number of related species. Most entries with many closely related species contain a key for differentiating among them. The example on page 10 shows the organization of a typical page and the recurrent features in each entry.

How to Use Identification Keys

You would probably not be holding this book if you didn't have at least a fleeting interest in learning the names of trees (though maybe you're just trying to impress someone). Either way, these pages contain a number of identification keys that satisfy an interest in identifying trees, and they are easy to use with a little practice. The keys begin by asking you to note a readily observable feature of the tree you wish to identify, and to select one of two (or rarely three) mutually exclusive choices about that feature. At each subsequent step in the key, on the path from unknown to known, you make a choice and, by process of elimination, advance toward a progressively smaller group of possibilities that match your unknown tree, until you identify the correct species. Using a key is like solving a puzzle and communicating with an expert (without the social awkwardness). It should be fun and empowering.

To progress through a key, you read both choices (usually written as a noun followed by one or more adjectives describing that noun), compare each statement with the tree being identified, and choose the statement that best matches your sample. For example, in the following key, which identifies three trees commonly found in California, you are asked whether the unknown tree has purple or green leaves. If the leaves are purple, the tree is a purple leaf plum; if the leaves are green, you are asked whether their undersides are white or green, and so on.

1. Leaves purple—Purple Leaf Plum (*Prunus cerasifera*)
1' Leaves green
    2. Leaf underside white—Olive (*Olea europaea*)
    2' Leaf underside green—Sweetgum (*Liquidambar styraciflua*)

The two mutually exclusive statements (called a couplet) always begin with the same number followed by a period or a prime symbol (') and have the same level of indentation. In some cases, the statement you choose will give you the name of the tree. If not, progress to the number directly below the chosen statement (not necessarily in numerical order) and continue making choices until you encounter a tree name. Always read both statements completely before deciding. It is best to examine more than one sample (not just one leaf, but ten leaves) to inform your decision. When measurements are given, use a ruler (one is provided on the back cover). The first time you use the keys in this book, you might practice with a tree whose name you already know.

Identification Key to Trees Commonly Cultivated in California

Trees denoted with (*) are not described in detail in the book.

1. Leaves >8 in. long, narrow, sword-shaped, simple (with undivided blades); leaf veins parallel—Group 1, Monocot trees
1' Leaves usually <8 in. long, not sword-shaped; leaves simple or compound; leaf veins parallel or netted
  2. Tree is a palm with large divided leaves arising from the summit of a single unbranched trunk—See the key to palm trees on page 165
  2' Tree is a conifer or broad-leaved tree, not a palm; leaves arise from branches
    3. Leaves needle-like, scale-like, or awl-like (very short, tightly pressed to the stem, hard, stiff, tapering to a point), or apparently absent—mostly conifers and a few angiosperms
      4. Leaves scale-like or awl-like, almost completely covering branchlets, mostly <½ in. long, or apparently absent—Group 2
      4' Leaves needle-like, ¾ in. long or longer—Group 3
    3' Leaf blades broad and flattened—mostly angiosperms
      5. Leaves compound (divided into leaflets)
        6. Leaves palmately compound—Group 4
        6' Leaves pinnately or bipinnately compound
          7. Leaves attached oppositely (2 per node)—Group 5
          7' Leaves attached alternately (1 per node)—Group 6
      5' Leaves simple (with blades not divided into leaflets, deeply lobed leaves are still simple)
        8. Leaves attached oppositely (2 per node) or whorled (3 or more per node)—Group 7
        8' Leaves attached alternately (1 per node)
          9. Leaf tip notched or truncate
            10. Leaves fan-shaped—Ginkgo (*Ginkgo biloba*)
            10' Leaves not fan-shaped
              11. Leaves with four or more pointed lobes, veins pinnate—Tulip Tree (*Liriodendron tulipifera*)
              11' Leaves with two rounded lobes, veins palmate
                12. Notch in leaf tip extending ⅓ down blade, flowers >1.5 in. wide, orchid-like—Orchid Trees (*Bauhinia* spp.)
                12' Notch in leaf tip <¼ in. or leaf without apical notch, flowers <1 in. wide, not orchid-like—Redbuds (*Cercis* spp.)
          9' Leaf tip pointed or rounded, not notched or truncate
            13. Leaf margin smooth (without teeth or lobes), wavy or flat—Group 8
            13' Leaf margin toothed or lobed
              14. Leaves palmately veined and/or palmately lobed (with veins and lobes radiating outward from a central point near the leaf base)—Group 9
              14' Leaves pinnately veined and/or pinnately lobed (with one main vein extending from the leaf base to the tip, smaller veins branching from that main vein, and lobes or teeth along the margins)
                15. Trees evergreen (leaves usually thick and leathery, found on stems of current and previous growing seasons)—Group 10
                15' Tree deciduous (leaves usually thinner, found only on stems of current growing season, leaf scars found on older stems, bud scale scars often present)—Group 11

Group 1—Trees with undivided, sword-shaped leaves, leaf veins parallel (monocotyledons)

1. Leaf margin with conspicuous teeth—Tree Aloe (*Aloe barberae* and *Aloe bainesii*)*
1' Leaf margin smooth or with minute, sharp teeth
    2. Leaf margin of all leaves smooth
        3. Leaves widest at the base; fruit orange; seeds 1 to 3—Dragon Tree (*Dracaena draco*)*
        3' Leaves widest at the middle; fruit blue or whitish; seeds numerous—Giant Dracaena, also called New Zealand Cabbage Palm (*Cordyline australis*)*
    2' Leaf margin with minute, sharp teeth (some occasionally smooth)
        4. Leaves <1 in. wide and >24 in. long—Pony Tail Palm (*Nolina recurvata*)*
        4' Leaves >1 in. wide, and if not >1 in. wide then <12 in. long
            5. Leaves 10 in. long or less—Joshua Tree (*Yucca brevifolia*)*
            5' Leaves 12 in. long or more—Giant Yucca (*Yucca elephantipes*)*

Group 2—Trees with scale-like, awl-like, or apparently absent leaves

1. Leaves apparently absent
    2. Tree thorny—Mexican Palo Verde (*Parkinsonia aculeata*)*
    2' Tree lacking thorns
        3. Branchlets ribbed, with whorls of minute leaves, lacking salt crystals—Sheoaks, also called Horsetail Trees (*Casuarina* spp.)
        3' Branchlets not ribbed, with a single, minute, scale-like leaf at each node, salt crystals often apparent—Athel Tree (*Tamarix aphylla*)
1' Leaves obvious, scale-like, awl-like, or occasionally needle-like
    4. Tree distinctly columnar
        5. Branchlets rope-like; leaves >¼ in. long, spreading; tree leaning—Cook Pine (*Araucaria columnaris*)
        5' Branchlets not rope-like; leaves <¼ in. long, tightly pressed to stem; tree not leaning— Italian Cypress (*Cupressus sempervirens*)
    4' Tree pyramidal, dome-shaped, tiered, or irregularly shaped, but not distinctly columnar
        6. Branchlets forming flat, two-dimensional sprays
            7. Leaves four per node—Incense Cedar (*Calocedrus decurrens*)
            7' Leaves two per node—Leyland Cypress (×*Hesperotropsis leylandii*) or Port Orford Cedar (*Chamaecyparis lawsoniana*)*
        6' Branchlets forming three-dimensional (not flat) sprays
            8. Leaves whorled in threes and/or opposite; seed cones fleshy, blue, berry-like— Junipers (*Juniperus* spp.)
            8' Leaves alternate or opposite, never whorled in threes; seed cones woody, brown
                9. Leaves >¼ in. long, spreading, awl-like
                    10. Bark fibrous, reddish-brown; seed cones <1.5 in. wide—Japanese Cryptomeria (*Cryptomeria japonica*)*
                    10' Bark smooth, gray, peeling, warty; seed cones >3 in. wide—Bunya Bunya, Norfolk Island Pine, and relatives (*Araucaria* spp.)
                9' Leaves <¼ in. long, tightly pressed to and overlapping on stem, awl- or scale-like
                    11. Leaves alternate, spiraling, awl-like—Giant Sequoia (*Sequoiadendron giganteum*)*
                    11' Leaves opposite, scale-like—Monterey Cypress (*Hesperocyparis macrocarpa*)

Group 3—Trees with needle-like leaves

1. Needle-like structures are actually jointed, ribbed branches with minute, whorled, scale-like leaves—Sheoaks, also called Horsetail Trees (*Casuarina* spp.)
1' Needle-like structures are unjointed leaves
   2. Leaves clustered in bundles
      3. Leaves in clusters of 2 to 5, each cluster surrounded at base by a sheath of scales—Pines (*Pinus* spp.)
      3' Leaves in dense clusters of 10 or more on short, stubby, lateral branchlets; sheath of scales absent—Cedars (*Cedrus* spp.)
   2' Leaves borne singly along stems, not clustered
      4. Leaves opposite—Dawn Redwood (*Metasequoia glyptostroboides*)
      4' Leaves alternate
         5. Bark smooth, gray, peeling, warty; seed cones >3 in. wide—Bunya Bunya, Cook Pine, Norfolk Island Pine and relatives (*Araucaria* spp.)
         5' Bark reddish brown to brown or gray, fibrous; seed cones <1.5 in. wide
            6. Foliage deciduous or partly so—Bald Cypresses (*Taxodium* spp.)*
            6' Foliage evergreen
               7. Leaves attached to branchlets with a distinct, short stalk
                  8. Leaves same color on both sides—Fern Pine (*Afrocarpus falcatus*)
                  8' Leaves distinctly lighter on underside
                     9. Leaves >¼ in. wide—Yew Pine (*Podocarpus macrophyllus*)
                     9' Leaves <¼ in. wide
                        10. Densely branched shrub; seed cones red, fleshy—Yews (*Taxus* spp.)*
                        10' Large tree; seed cones brown, woody—Douglas Firs (*Pseudotsuga* spp.)*
               7' Leaves sessile or tapering only slightly to branchlets
                  11. Leaves with a distinct keel, triangular in cross section—Japanese Cryptomeria (*Cryptomeria japonica*)*
                  11' Leaves more or less flat—Coast Redwood (*Sequoia sempervirens*)

Group 4—Trees with palmately compound leaves

1. Leaflets mostly 3
   2. Leaves alternate
      3. Leaflets >¾ in. wide, stalked—Coral Trees (*Erythrina* spp.)
      3' Leaflets ½ in. wide or less, sessile—African Sumac (*Searsia lancea*)
   2' Leaves opposite or whorled
      4. Leaflet margin smooth—Puriri (*Vitex lucens*)*
      4' Leaflet margin irregularly toothed or lobed—Box Elder (*Acer negundo*)
1' Leaflets usually 5 or more
   5. Leaves alternate
      6. Leaflet margin serrate—Floss Silk Tree (*Ceiba speciosa*)
      6' Leaflet margin smooth—Umbrella Tree (*Schefflera actinophylla*)*
   5' Leaves opposite or whorled
      7. Leaflet margin smooth, leaves covered with golden hairs—Golden Trumpet Tree (*Handroanthus chrysotrichus*)
      7' Leaflet margin toothed, leaves hairless or not
         8. Leaflets sessile
            9. Flowers red—Red Horse Chestnut (*Aesculus* × *carnea*)
            9' Flowers white—European Horse Chestnut (*Aesculus hippocastanum*)*

8′ Leaflets stalked

    10. Leaflet stalks all similar in length; fruit spherical; tree deciduous in winter—California Buckeye (*Aesculus californica*)*

    10′ Leaflet stalks of differing lengths; fruit a linear, narrow capsule; tree with leaves in winter—Pink Trumpet Tree (*Handroanthus heptaphyllus*)

Group 5—Trees with pinnately and bipinnately compound, opposite leaves

1. Leaves bipinnately compound (twice divided)

    2. Leaves with >10 pairs of primary leaflets—Jacaranda (*Jacaranda mimosifolia*)

    2′ Leaves with <8 pairs of primary leaflets—Catalina Ironwood (*Lyonothamnus floribundus* subsp. *aspleniifolius*)

1′ Leaves pinnately compound (once divided)

    3. Leaflets 5 or fewer

        4. Leaflet margin smooth—Puriri (*Vitex lucens*)*

        4′ Leaflet margin irregularly toothed or lobed—Box Elder (*Acer negundo*)

    3′ Leaflets 7 or more

        5. Leaflet stalks >¼ in. long—Ashes (*Fraxinus* spp.)

        5′ Leaflets sessile or with stalks <¼ in.

            6. Leaflet margin divided or lobed—Catalina Ironwood (*Lyonothamnus floribundus* subsp. *aspleniifolius*)

            6′ Leaflet margin smooth, flat, or wavy

                7. Leaflet tip rounded or slightly notched—Tipu Tree (*Tipuana tipu*)

                7′ Leaflet tip pointed or tapering to a point

                    8. Leaflets <¾ in. wide—Raywood Ash (*Fraxinus angustifolia* 'Raywood')

                    8′ Leaflets >1 in. wide

                        9. Central leaf axis covered with rusty hairs; flowers red and orange—African Tulip Tree (*Spathodea campanulata*)

                        9′ Central leaf axis without rusty hairs; flowers yellow with red stripes—Markhamia (*Markhamia lutea*)*

Group 6—Trees with pinnately and bipinnately compound, alternate leaves

1. Leaflets 3

    2. Leaflets >¾ in. wide, stalked—Coral Trees (*Erythrina* spp.)

    2′ Leaflets <½ in. wide, sessile—African Sumac (*Searsia lancea*)

1′ Leaflets 5 or more

    3. Leaves bipinnately compound (twice divided)

        4. Tree with thorns or spines

            5. Thorns single; central leaf axis flattened, and often forked—Mexican Palo Verde (*Parkinsonia aculeata*)*

            5′ Thorns branched; central leaf axis round in cross section, not forked—Honey Locust (*Gleditsia triacanthos*)

        4′ Tree lacking thorns or spines

            6. Leaves with raised spherical glands on upper surface of central leaf axis; flowers yellow, clustered in spherical heads or spikes—Acacias (*Acacia* spp.)

            6′ Leaves without glands on upper surface; flowers varying

                7. Leaflet margin smooth or only slightly toothed

                    8. Leaflet bases symmetrical; flowers green; fruit >1 in. wide—Honey Locust (*Gleditsia triacanthos*)

                    8′ Leaflet bases distinctly asymmetrical; flowers pink; fruit <¾ in. wide—Silk Tree (*Albizia julibrissin*)

7' Leaflet margin conspicuously serrate, irregularly toothed, or lobed

    9. Leaflet undersides hairy and silver—Silk Oak (*Grevillea robusta*)

    9' Leaflet undersides hairless

        10. Bark rough, dark brown; tree deciduous; flowers purple; fruit not inflated—Chinaberry (*Melia azedarach*)

        10' Bark smooth, gray; tree evergreen or partly deciduous; flowers yellow; fruit inflated—Chinese Flame and Flamegold Trees (*Koelreuteria* spp.)

3' Leaves pinnately compound (once divided)

    11. Leaves with a pair of leaflets at the tip

        12. Leaflets <1.5 in. long—Honey Locust (*Gleditsia triacanthos*)

        12' Leaflets 1.5 in. long or longer

            13. Leaflets <½ in. wide—Peruvian Pepper Tree (*Schinus molle*)

            13' Leaflets ¾ in. wide or greater

                14. Leaflets with a rounded or notched tip

                    15. Leaflets <2 times longer than wide; margins flat; fruit cylindrical—Carob Tree (*Ceratonia siliqua*)

                    15' Leaflets 2 times longer than wide; margins rolled under; fruit spherical—Carrotwood (*Cupaniopsis anacardioides*)

                14' Leaflet with a pointed tip

                    16. Leaflet margin toothed or occasionally smooth—Chinese Cedrela (*Cedrela sinensis*)*

                    16' Leaflet margin always smooth

                        17. Leaflets stalked; flowers large and golden; tree evergreen; fruit >6 in. long, square in cross section—Gold Medallion Tree (*Cassia leptophylla*)

                        17' Leaflets sessile or nearly so; flowers inconspicuous; tree deciduous; fruit spherical, <½ in. wide—Chinese Pistache (*Pistacia chinensis*)

    11' Leaves with a single terminal leaflet

        18. Leaflet margin toothed or lobed

            19. Leaflet undersides hairy and silver—Silk Oak (*Grevillea robusta*)

            19' Leaflet undersides green, hairless

                20. Leaflets <½ in. wide

                    21. Leaflet tip tapering to a point; flowers white; fruit spherical, red—Peruvian Pepper Tree (*Schinus molle*)

                    21' Leaflet tip rounded to a point; flowers green; fruit a strap-like legume, brown—Honey Locust (*Gleditsia triacanthos*)

                20' Leaflets ¾ in. wide or greater

                  22. Leaflets coarsely toothed, doubly serrate with small teeth on larger teeth—Goldenrain Tree (*Koelreuteria paniculata*)

                  22' Leaflets evenly toothed, singly serrate

                    23. Leaflets farthest from the leaf base largest; fruit four-sided, splitting at maturity—Pecan (*Carya illinoinensis*)*

                    23' Leaflets uniform in size or middle ones largest; fruit round, not splitting—Walnuts (*Juglans* spp.)

        18' Leaflet margin smooth

            24. Leaflets stalked

                25. Leaflet tip tapering to a point

                    26. Mature leaves mostly >12 in. long—Tree of Heaven (*Ailanthus altissima*)

                    26' Mature leaves <12 in. long—Japanese Pagoda Tree (*Styphnolobium japonica*)

                25' Leaflet tip rounded or rounded to a point and/or notched

27. Leaflet leathery; tree evergreen; most leaflet tips notched—Carrotwood (*Cupaniopsis anacardioides*)

27' Leaflet papery (not leathery); tree deciduous; most leaflet tips not notched

    28. Leaves with >20 leaflets; flowers yellow; fruit winged—Tipu Tree (*Tipuana tipu*)

    28' Leaves with <20 leaflets; flowers white or pink; fruit not winged—Black Locust (*Robinia pseudoacacia*)

24' Leaflets sessile or nearly so

  29. Leaflet tip rounded and usually notched—Carrotwood (*Cupaniopsis anacardioides*)

  29' Leaflet tip pointed, not notched

    30. Leaflets ¼ in. wide or less—Peruvian Pepper Tree (*Schinus molle*)

    30' Leaflets ½ in. wide or greater

      31. Leaflets asymmetrical, sickle-shaped—South African Plum (*Harpephyllum caffrum*)*

      31' Leaflets symmetrical, egg- or spearhead-shaped

        32. Leaflets >1.5 in. wide—English Walnut (*Juglans regia*)

        32' Leaflets <1.5 in. wide

          33. Leaflet tip rounded to a point; central leaf axis distinctly winged; leaflets leathery; tree evergreen; flowers with petals—Brazilian Pepper Tree (*Schinus terebinthifolius*)

          33' Leaflet tip tapering to a point; central leaf axis not winged; leaflets papery; tree deciduous; flowers without petals—Chinese Pistache (*Pistacia chinensis*)

Group 7—Trees with simple, opposite, or whorled leaves

1. Bark thick, papery, and peeling; flower clusters resembling bottle brushes; clusters of sessile woody fruits persisting along stems—Melaleuca (*Melaleuca* spp.). Only some species have opposite leaves: see *Melaleuca* treatment.

1' Tree without the set of characteristics described above

  2. Leaves whorled, or both whorled and opposite on same tree

    3. Leaf underside woolly white—Pineapple Guava (*Feijoa sellowiana*)*

    3' Leaf underside green

      4. Leaf margin toothed—Macadamia nut (*Macadamia integrifolia*)*

      4' Leaf margin smooth—Oleander (*Nerium oleander*)*

  2' Leaves always opposite

    5. Leaves >8 in. wide

      6. Leaves mostly egg-shaped; fruit a long cylindrical capsule; flowers white or yellow—Catalpa (*Catalpa* spp.)

      6' Leaves mostly heart-shaped; fruit an egg-shaped capsule; flowers purple—Empress Trees (*Paulownia tomentosa* [peduncles as long or longer than pedicels] or *Paulownia kawakamii* [peduncles much shorter than pedicels])*

    5' Leaves <6 in. wide

      7. Leaves palmately veined

        8. Leaves heart-shaped, margin with rounded teeth; fruit a follicle—Katsura Tree (*Cercidiphyllum japonicum*)*

        8' Leaves not heart-shaped, margin lobed and often with pointed teeth, or occasionally smooth; fruit with two wings—Maples (*Acer* spp.)

      7' Leaves pinnately veined

        9. Leaves bluish green or bluish gray (glaucous); petals and sepals fused into a bud cap—Eucalypts, also called Gum Trees (*Eucalyptus* spp.)

        9' Leaves green

          10. Leaves sessile or nearly so

11. Leaves ¾ in. wide or wider; new branches square in cross sections—Crape Myrtle (*Lagerstroemia* spp.)

11′ Leaves <½ in. wide; branch tips round in cross section—Desert Willow (*Chilopsis linearis*)*

10′ Leaves clearly stalked

12. Leaves ¼ in. wide and linear (8 times longer than wide)—Desert Willow (*Chilopsis linearis*)*

12′ Leaves ½ in. wide or wider, not linear

13. Leaf undersides white and woolly

14. Leaves egg-shaped, with rounded tip—New Zealand Christmas Tree (*Metrosideros excelsa*)

14′ Leaves shaped like spearheads, or oval-shaped, with pointed tip—Olive (*Olea europaea*)

13′ Leaf undersides green, light green, or gray (occasionally hairy, but not white and woolly)

15. Secondary leaf veins curving upward from the midvein to the tip; when leaves are torn in half, vein strings are visible between pieces; flowers subtended by showy colorful or white bracts—Dogwoods (*Cornus* spp.)

15′ Most secondary leaf veins parallel; no vein strings visible between torn leaf halves; showy bracts absent

16. New stem growth square in cross section, often with two fleshy protrusions above each node; young leaves reddish—Brush Cherry (*Syzygium australe*)

16′ New stem growth round in cross section, no protrusions above nodes; young leaves green or occasionally red

17. Leaves shaped like spearheads

18. Leaves thin, papery; tree deciduous—Desert Willow (x*Chitalpa tashkentensis*)

18′ Leaves thick, leathery; tree evergreen, Eucalyptus-like— Rose Gum (*Angophora costata*)*

17′ Leaves egg-shaped, oval, or long oval

19. Leaf stalk and base of midvein on underside of leaf hairy; tree deciduous; bark dark brown and rough—Chinese Fringe Tree (*Chionanthus retusus*)

19′ Leaf stalk and underside hairless; tree evergreen; bark gray and smooth

20. Leaf stalks and branchlets gray and green; leaves with translucent dots, with >10 pairs of lateral veins; flowers pink purple—Cape Chestnut (*Calodendrum capense*)*

20′ Leaf stalks and branchlets reddish brown; leaves lacking dots, with <5 pairs of lateral veins; flowers white— Privets (*Ligustrum lucidum*, leaves 4–6 in. long; *Ligustrum japonicum*, leaves 2–4 in. long)

Group 8—Trees with alternate, simple, smooth leaves

1. Flower clusters resembling bottle brushes; clusters of sessile, cup-shaped, woody fruits persisting along stems

2. Flowers mostly red, stamens not clustered into groups—Bottlebrushes (*Callistemon* spp.)

2′ Flowers mostly white or pink, stamens clustered in 5 groups—Melaleucas (*Melaleuca* spp.)

1′ Flower clusters not resembling bottle brushes; tree without clusters of sessile, woody fruits persisting along stems

3. Leaves are phyllodes (leaf-like structures derived from the flattening of the leaf stalk and central leaf axis), mostly <1 in. wide, each attached to the stem at a swelling (pulvinus), and with an inconspicuous yellow gland ⅛ in. to ½ in. away from the node on the upper phyllode margin; flowers cream-colored to bright yellow; fruit a leathery legume—Acacias (*Acacia* spp.)

3′ Leaves not phyllodes; tree without the combination of characters described above

  4. Most leaves ½ in. wide or less

    5. Leaves <1 in. long

      6. Leaf tips spiny, leaves sessile; tree a large conifer—Bunya Bunya (*Araucaria bidwillii*) or Monkey Puzzle (*A. araucana*)

      6′ Leaf tips rounded with a small nipple, leaves stalked; tree small, flowering—Tea Trees (*Leptospermum laevigatum** or, if leaves needle-like, flowers pink or red, *Leptospermum scoparium*)*

    5′ Leaves >1.25 in. long

      7. Leaves lemon-scented when crushed; bark smooth and cream-colored—Lemon Scented Gum (*Corymbia citriodora*)

      7′ Leaves unscented or smelling medicinal or spicy when crushed (but not lemon-scented); bark varying

        8. Branch ends thorny—Simple-Leaved Pepper Tree (*Schinus polygamus*)*

        8′ Branch ends not thorny

          9. Leaves with translucent dots when held to the light

            10. Flowers and fruit at branch ends; no leaf vein parallel to leaf margin; bark shallowly furrowed and gray—Australian Willow (*Geijera parviflora*)

            10′ Flowers and fruit in leaf axils; leaf vein parallel to leaf margin conspicuous; bark varying

              11. Bark smooth, flaking, white or gray; flowers yellow—Water Gum (*Tristaniopsis laurina*)

              11′ Bark fibrous brown; flowers white

                12. Petals fused into a green bud cap; leaves bluish green; branchlets straight; mature trees >40 ft.—Narrow-Leaf Peppermint (*Eucalyptus nicholii*)

                12′ Petals white, conspicuous; leaves olive green; branchlets zigzagged; mature trees 20 to 40 ft.—Willow Peppermint (*Agonis flexuosa*)

          9′ Leaves without translucent dots

            13. Mature leaves mostly <2 in. long—Fern Pine (*Afrocarpus falcatus*)

            13′ Mature leaves mostly >2.5 in. long

              14. Leaves reddish purple or bronze, occasionally sticky, glandular, margins rolled downward—Hopseed (*Dodonaea viscosa*)*

              14′ Leaves green, not glandular, margins flat

                15. Leaf midrib extending beyond blade, ending in a short apical hook; branchlets drooping; fruit an orange capsule—Narrow-Leaf Pittosporum (*Pittosporum angustifolium*)

                15′ Leaf midrib not extending beyond blade; branchlets upright; fruit-like seed blue with a red, fleshy attachment—Yew Pine (*Podocarpus macrophyllus*)

  4′ Most leaves ¾ in. wide or greater

    16. Tree bleeding milky white sap when damaged (when a leaf is pulled off)

      17. Tree deciduous; leaves with two swollen glands at attachment of leaf blade and leaf stalk—Chinese Tallow Tree (*Triadica sebifera*)

      17′ Tree evergreen; leaves without glands—Figs (*Ficus* spp.)

    16′ Tree without milky white sap

18. Leaves lemon-scented when crushed; bark smooth and cream-colored—Lemon Scented Gum (*Corymbia citriodora*)

18' Leaves unscented or smelling medicinal or spicy when crushed (but not lemon-scented); bark varying

    19. Leaves with translucent dots when held to the light

        20. Petals fused into a bud cap; fruit a woody capsule—Eucalypts, also called Gum Trees (*Eucalyptus* spp.)

        20' Petals conspicuous, not fused

            21. Bark smooth, peeling away in irregular strips and plates; flowers yellow; fruit brown, woody—Water Gum (*Tristaniopsis laurina*)

            21' Bark smooth or rough, not peeling away; flowers white; fruit fleshy

                22. Leaf stalk winged or margined, jointed at base of blade; fruit >2 in. wide, orange to yellow—Citrus (*Citrus* spp.)*

                22' Leaf stalk not winged; fruit <1 in. wide, red—Myoporum (*Myoporum laetum*)

    19' Leaves without translucent dots

        23. Leaves heart-shaped—Eastern Redbud (*Cercis canadensis*)

        23' Leaves not heart-shaped

            24. Leaf undersides white, yellowish-white, scaly or woolly (but not rust-colored)

                25. Margins of some leaves toothed; fruit an acorn; bark dark gray to black—Holly Oak (*Quercus ilex*)

                25' Margins of all leaves smooth; fruit a capsule; bark varying

                    26. Leaf underside scaly; seeds reddish brown, not sticky—Primrose Tree (*Lagunaria patersonia*)

                    26' Leaf underside woolly; seeds black, sticky—Karo (*Pittosporum crassifolium*)

            24' Leaf underside green, light green, or rust-colored, occasionally sparsely hairy (but not white, yellowish-white, woolly, or scaly)

                27. Leaf underside with rust-colored hairs (at least when young), upper surface dark green, shiny

                    28. Flowers at branch ends; leaf undersides light green and rusty—Southern Magnolia (*Magnolia grandiflora*)

                    28' Flowers in leaf axils; leaf undersides pale waxy, bluish white and rusty—Sweet Michelia (*Michelia doltsopa*)

                27' Leaf underside green, light green, or gray, upper surface dull or shiny

                    29. Leaf upper surface slightly rough to the touch, dull, with minute hairs; flowers >2 in. wide—Saucer Magnolia (*Magnolia* × *soulangeana*)

                    29' Leaf upper surface smooth to the touch, shiny, hairless; flower <1.5 in. wide

                        30. Leaf stalks and branchlets with fine hairs; flowers white to yellow—Sweetshade (*Hymenosporum flavum*)

                        30' Leaf stalks, young leaves, and branchlets hairless; flowers varying in color

                            31. Leaf stalk swollen at tip where attached to leaf blade, leaf margins variously lobed or smooth, occasionally deeply palmately lobed; fruit a woody canoe-like follicle—Bottle Trees (*Brachychiton* spp.)

                            31' Leaf stalk without swelling at the tip, leaf margins always smooth or wavy; fruit varying

                                32. Leaves with 3 main veins converging at 2 glands, smelling of camphor when crushed—Camphor Tree (*Cinnamomum camphora*)

32' Leaves with 1 main central vein, not smelling of camphor when crushed
    33. Leaves with two swollen glands at attachment of leaf blade and leaf stalk—Chinese Tallow Tree (*Triadica sebifera*)
    33' Leaves without swollen glands
        34. Leaf margin wavy with regular undulations; fruit an orange capsule; flowers white—Pittosporums (*Pittosporum* spp.)
        34' Leaf margin flat or slightly and irregularly wavy; fruit fleshy or a brown, woody capsule
            35. Leaves sessile or leaf stalk <⅛ in., young stems square—Crape Myrtle (*Lagerstroemia* spp.)
            35' Leaf stalk >⅛ in., young stems square or round in cross section
                36. Fruit a brown, woody capsule; tree Eucalyptus-like
                    37. Fruit ¾ in. wide or wider; flowers red, orange, or pink—Red Flowering Gum (*Corymbia ficifolia*)
                    37' Fruit <½ in. wide; flower variously colored
                        38. Petals fused into a bud cap—Eucalypts, also called Gum Trees (*Eucalyptus* spp.)
                        38' Petals not fused
                            39. Petals white; bark cinnamon-brown—Brisbane Box (*Lophostemon confertus*)
                            39' Petals yellow; bark gray or white—Water Gum (*Tristaniopsis laurina*)
                36' Fruit fleshy, black, blue or green; tree not Eucalyptus-like
                    40. Leaves strongly scented, smelling spicy when crushed (bay leaves), margins yellow, translucent when held to the light; tree evergreen, often multi-stemmed—Grecian Laurel or Sweet Bay (*Laurus nobilis*)
                    40' Leaves not strongly scented, margins opaque; tree evergreen or deciduous
                        41. Leaves thin, deciduous; fruit <½ in. wide
                            42. Leaf veins pinnate, with one main vein—Sour Gum or Tupelo (*Nyssa sylvatica*)*
                            42' Leaf veins palmate with 3 or more veins radiating outward from a central point near the leaf base—Chinese Hackberry (*Celtis sinensis*)
                        41' Leaves thick, leathery, evergreen; fruit >½ in. wide
                            43. Leaves 1 in. wide or less; fruit <1.5 in. wide—Cherry Laurels (*Prunus caroliniana, P. laurocerasus,* occasionally *P. ilicifolia* leaves are smooth-margined)*
                            43' Leaves >1.5 in. wide, fruit >2 in. wide—Avocado (*Persea americana*)*

Group 9—Trees with alternate, simple, palmately veined and/or palmately lobed leaves

1. Tree bleeding milky white sap when damaged (when a leaf is pulled off); deciduous stipules forming a ring-like scar around each node
    2. Leaves deeply 3- to 5-lobed; flowers inconspicuous, borne inside figs—Edible Fig (*Ficus carica*)
    2' Leaves irregularly lobed, or some not lobed and with toothed margins; flowers in spikes—White Mulberry (*Morus alba*)
1' Tree not bleeding milky white sap; stipular ring scar present or not
    3. Leaves not palmately lobed and leaf margins toothed

4. Leaf underside hairy, at least along veins, or bluish green with wax (glaucous)—Linden Trees (*Tilia* spp.)

4′ Leaf underside hairless

    5. Leaves triangle-shaped—Lombardy Poplar (*Populus nigra* 'Italica')

    5′ Leaves egg- or spearhead-shaped

        6. Leaf veins conspicuously raised on underside; flowers blue or purple; fruit green or brown—Wild Lilacs, also called California Lilacs (*Ceanothus* spp. and varieties; *Ceanothus* 'Ray Hartman' is widely grown)

        6′ Leaf veins not conspicuously raised on underside; flowers white; fruit red—Lavalle Hawthorn (*Crataegus* × *lavallei*)

3′ Leaves palmately lobed and lobes toothed or lobes smooth

    7. Mature leaves <3 in. wide; trees occasionally thorny—Hawthorns (leaf blade tapering to leaf stalk—English Hawthorn [*Crataegus laevigata*]; base of leaf blade truncate—Washington Hawthorn [*Crataegus phaenopyrum*])

    7′ Mature leaves 3.5 in. wide or greater, trees never thorny

        8. Base of leaf stalk swollen, covering axillary bud, stipules encircling the stem and leaving a ring scar after falling—Plane Trees and Sycamores (*Platanus* spp.)

        8′ Base of leaf stalk not swollen, stipules free, no ring scar around nodes

            9. Leaves spicy when crushed, lobe margins regularly sawtoothed

                10. Leaves mostly 5-pointed—Sweetgum (*Liquidambar styraciflua*)

                10′ Leaves mostly 3-pointed—Formosan Sweetgum (*Liquidambar formosana*)

            9′ Leaves unscented even when crushed, lobe margins smooth or irregularly toothed

                11. Leaf bases heart-shaped, lobes often overlapping, flowers >2 in. wide, upright—Monkey Hand Tree (*Chiranthodendron pentadactylon*)*

                11′ Leaf bases tapering, rounded, or truncated at leaf stalk, basal lobes if present not overlapping; flowers <1.5 in. wide, facing downward

                    12. Fruit a membranous capsule, green, splitting open before completely developed—Chinese Parasol Tree (*Firmania simplex*)*

                    12′ Fruit a woody follicle, brown, not splitting until developed—Bottle Trees (*Brachychiton* spp.)

Group 10—Evergreen trees with alternate, simple, pinnately veined and/or pinnately lobed leaves

1. Fruit an acorn—Oaks (*Quercus* spp.)

1′ Fruit not an acorn

    2. Leaves with translucent dots when held to the light

        3. Leaf stalks winged or margined, jointed at base of blade; fruit >2 in. wide, orange to yellow—Citrus (*Citrus* spp.)*

        3′ Leaf stalks not winged; fruit <1 in. wide, red—Myoporum (*Myoporum laetum*)

    2′ Leaves without translucent dots

        4. Leaves <½ in. wide; tree form weeping—Mayten Tree (*Maytenus boaria*)

        4′ Leaves >½ in. wide or greater; tree form varying

            5. Leaf margin wavy (undulate), spiny, prickly, or variously lobed

                6. Leaf margin sawtoothed, smooth, or variously lobed all on the same tree; flowers orange or red in a wheel-shaped cluster; fruit woody, brown—Firewheel Tree (*Stenocarpus sinuatus*)

                6′ Leaf margin spiny and prickly; flowers white or greenish, clusters not wheel-like

                    7. Marginal spines coarse, fewer than 20—Hollies (*Ilex* spp.)*

                    7′ Marginal spines slender, more than 20—Hollyleaf Cherry (*Prunus ilicifolia*) and Portugal Laurel (*P. lusitanica*)*

            5′ Leaf margin evenly sawtoothed (not spiny), flat or wavy

8. Leaf blades hairy, at least on one surface, especially on new growth
   9. Leaves <3 in. long—Toyon (*Heteromeles arbutifolia*)
   9' Leaves >4 in. long—Loquats: underside hairless (glabrous)—Bronze Loquat (*Eriobotrya deflexa*); leaf underside hairy (woolly)—Edible Loquat (*Eriobotrya japonica*)
8' Leaf blades hairless on both surfaces
   10. Young stems and occasionally leaf stalks with conspicuous glandular hairs—Strawberry Trees (single-trunked, bark smooth, reddish brown—*Arbutus* 'Marina'; multi-trunked, bark rough, shed in small cracks—*Arbutus unedo*)
   10' Young stems and leaf stalks hairless or occasionally with fine, non-glandular hairs
      11. Leaf stalks >½ in. long
         12. Leaves egg-shaped, new growth green—Evergreen Pear (*Pyrus kawakamii*)
         12' Leaves oblong, new growth red—Chinese Photinia (*Photinia serrulata*)*
      11' Leaf stalks <½ in. long
         13. Leaves <¾ in. wide; bark shedding in small plates, with orange patches—Chinese Elm (*Ulmus parvifolia*)
         13' Leaves >1 in. wide; bark varying
            14. Margins of mature leaves red—Xylosma (*Xylosma congestum*)*
            14' Margins of mature leaves green
               15. Leaf undersides bluish green, some leaves opposite or almost so; flowers pendent, solitary; fruit dry—Lily of the Valley Tree (*Crinodendron patagua*)*
               15' Leaf undersides green, all leaves alternate; flowers formed in upright spikes; fruit fleshy—Hollyleaf Cherry (*Prunus ilicifolia*) and Portugal Laurel (*P. lusitanica*)*

Group 11—Deciduous trees with alternate, simple, pinnately veined and/or pinnately lobed leaves

1. Fruit an acorn—Oaks (*Quercus* spp.)
1' Fruit not an acorn
   2. Leaf bases asymmetrical, sometimes only slightly so
      3. Leaf margin with silky hairs—European Beech (*Fagus sylvatica*)*
      3' Leaf margin without silky hairs
         4. Leaves with 3 major veins spreading from base; bark smooth or with irregular corky warts and/or ridges; fruit fleshy—Hackberries (*Celtis* spp.)
         4' Leaves with 1 main central vein; bark scaly or fissured into parallel ridges, without warty protrusions; fruit dry
            5. Bark scaly, with orange and gray patches; leaf margins with bristle-tipped teeth; fruit not winged—Sawleaf Zelkova (*Zelkova serrata*)
            5' Bark fissured, brown; leaf margins sawtoothed (often with small teeth on larger teeth), teeth not bristle-tipped; fruit winged—Elms (*Ulmus* spp.)
   2' Leaf bases symmetrical
      6. Trees dioecious, with a single bud-scale leaf; seeds attached to silky hairs—Willows (*Salix* spp.)*
      6' Trees monoecious or flowers bisexual, with multiple bud-scale leaves; seeds without silky hairs; not a willow
         7. Fruit fleshy, colorful
            8. Fruit a one-seeded drupe, without floral remnants on the end—Flowering Cherries and Plums (*Prunus* spp.)
            8' Fruit a multi-seeded pome, the tip with obvious 5-parted floral remnants on the end

9. Flowers clustered on branched inflorescences; fruit with 1 to 5 bony nuts—Hawthorns (*Crataegus* spp.)

9' Flowers solitary; fruit divided into 1 to 5 chambers, each with 1 or more seeds

    10. Fruit without gritty cells, red, yellow, or purple; styles joined at the base—Flowering Crab Apples (*Malus* spp.)

    10' Fruit with gritty cells, brown or tan; styles free

        11. Tree deciduous for >1 month; bark gray, shallowly fissured, often with a compact branching structure at the top of the trunk—Callery or Bradford Pear (*Pyrus calleryana*)

        11' Tree deciduous only for a short period or evergreen; bark dark gray to black, deeply checkered; tree with an open, branching structure—Evergreen Pear (*Pyrus kawakamii*)

7' Fruit dry, woody, brown

    12. Leaf margin wavy, with teeth only above the middle—Persian Parrotia (*Parrotia persica*)*

    12' Leaf margin flat, regularly sawtoothed throughout

        13. Leaves rough to the touch, margins with bristle-tipped teeth—Sawleaf Zelkova (*Zelkova serrata*)

        13' Leaves not rough to the touch, marginal teeth not bristle-tipped

            14. Bark smooth, mottled, with orange patches, shed in plates—Chinese Elm (*Ulmus parvifolia*)

            14' Bark rough, fibrous, dark brown to black, or smooth and gray or white

                15. Fruits are tiny, flattened, winged nuts housed in cone-like structures

                    16. Cones on branched stems, persistent, woody—Alders (*Alnus* spp.)

                    16' Cones solitary, breaking apart at maturity—Birches (*Betula* spp.)

                15' Fruits are large nuts, not in cone-like structures, not flattened or winged

                    17. Male flowers in erect clusters; fruit attached to a leafy bract—European Hornbeam (*Carpinus betulus*)*

                    17' Male flowers hanging in catkins; fruit a nut enclosed in a scale-like bract—Hazels (*Corylus* spp.)*

Compendium of Trees

California
Valley Oak

Gymnosperms

Afrocarpus falcatus   Fern Pine

AFF-roe-kar-pus fall-KAY-tuss
Afro - Gr., African; *karpos* - a fruit
falcatus - L., sickle-shaped
Podocarpaceae

Uganda, Ethiopia, and Kenya
Simple, Alternate
Evergreen, 50–60 ft.

Synonyms: *Podocarpus falcatus,* and often misidentified as *Podocarpus gracilior*

The 150 or so species in the strange and diverse yellowwood family (Podocarpaceae) grow mainly in the Southern Hemisphere. The African fern pine, one of relatively few conifers native to Africa, was recently separated from the genus *Podocarpus* into the genus *Afrocarpus*. *Afrocarpus* cones lack the fleshy, often brightly colored, structures (called receptacles) of *Podocarpus* cones. Despite what is described in many books, this species does not produce flowers or fruits; it is a conifer, but recurring confusion arises because instead of producing woody cones like most conifers, the fern pines produce fleshy, greenish yellow, plumlike cones. The fern pine is the most commonly cultivated of the Podocarpaceae. Besides the distinctively fleshy cones, it can be recognized by its rounded, evergreen crown of dark green, linear leaves and its thin, scaly, pale gray bark. The fern pine is deer- and drought-resistant, grows well in poor soils, ocean winds, air pollution, and urban environments. The yew pine (*P. macrophyllus*) and the long-leaved yellowwood (*P. henkelii*) are also regularly grown in California.

A. falcatus linear leaves
and old male cones

A. falcatus fleshy cones
and linear leaves

P. macrophyllus cones, each
with a fleshy, red *receptacle*

*In the movement of trees, I find my
own agitation. —Wallace Stevens*

P. macrophyllus   *A. falcatus*   *A. falcatus* bark

Araucaria bidwillii   Bunya Bunya

ar-uh-KAIR-ee-ah bid-WILL-ee-eye

Araucanos - Chilean for monkey puzzle tree

bidwillii - John Carne Bidwill (1815–1853)

Araucariaceae

Queensland, Australia

Simple, Alternate

Evergreen, 100 ft.

The distinctive and ancient conifer family Araucariaceae (ar-uh-KAIR-ee-aye-see-ee) is made up of only three genera, *Agathis* (21 species), *Araucaria* (19 species), and *Wollemia*, with a single species: *W. nobilis* (Wollemi pine). *Araucaria* is the most commonly cultivated genus in the family and two species are widespread in California. The bunya bunya is a prehistoric-looking, imposing tree with regular, spiral, wide-sweeping branches with dark green, leathery leaves. Most trees produce both small, cylindrical male cones and enormous, pineapple-like female cones that fall apart when mature, usually while still attached to the tree. Occasionally one of these bowling-ball-sized, ten-pound cones will fall from a great height, leaving a large divot in the lawn where it lands. It's best not to stand under a bunya bunya on a windy day!

Cone scale with single
edible seed

*The highest and most
lofty trees have the
most reason to dread
the thunder.*
—*Charles Rollin*

Sharp leaves

Female cone

ar-uh-KAIR-ee-ah het-er-AWE-fill-uh

Araucanos - Chilean for monkey puzzle tree

hetero - Gr., different; *phyllon* - Gr., leaf

Araucariaceae

Norfolk Islands

Simple, Alternate

Evergreen, 150 ft.

A. heterophylla

A. columnaris

The Norfolk Island pine, which is not a true pine (*Pinus* spp.), is the most commonly cultivated *Araucaria* in California, although Cook pine (*A. columnaris*) is more commonly grown worldwide. Young *Araucaria* are regularly sold as potted houseplants and grow well into mature trees in mild coastal climates. This tree has an unmistakably symmetrical, pyramid-like form with horizontal, layered branches upturned at the tips that form a starlike arrangement from a perfectly straight trunk. The silhouettes of these tall trees, with this characteristic branching, can be easily recognized up and down the California coast. Conservatively estimated, the number of tiny leaves on a mature 80-year-old, 100-foot-tall tree is around fifty million! The papery thin outer bark is smooth, gray-brown, and studded with black protrusions and curls off the trunk in small sections.

If a tree dies, plant another in its place. —Carl Linnaeus

Bark

Male cones

Female cone and leaves

Key to California's Commonly Cultivated *Araucaria*

1. Leaves ¾ in. long or longer
    2. Leaves spreading, often arranged in two rows, narrowing at the base, stiff but flexible—Bunya Bunya (*A. bidwillii*)
    2′ Leaves strongly overlapping, often spirally arranged, barely narrowing at the base, stiff and rigid—Monkey Puzzle (*A. araucana*)
1′ Leaves ½ in. long or shorter
    3. Crown of mature tree rounded, upper branches arching upward, foliage in dense tufts at ends of branches—Hoop Pine (*A. cunninghamii*)
    3′ Crown of mature tree pointed, upper branches angled, foliage distributed throughout length of branches
        4. Mature tree columnar, often leaning—Cook Pine (*A. columnaris*)
        4′ Mature tree conical, not leaning—Norfolk Island Pine (*A. heterophylla*)

Calocedrus decurrens   Incense Cedar

kal-owe-SEE-druss deh-KURR-enz

Kallos - Gr., beautiful; *kedros* - Gr., cedar

decurrens - L., running down the stem

Cupressaceae

California, Nevada, Oregon, Baja California

Scale-like, Opposite

Evergreen, 60–90 ft.

The incense cedar grows natively in mixed coniferous forests from northern Baja California to the southern slopes of Oregon's Mt. Hood and occurs most frequently above 2,000 foot elevation in the Sierra Nevada. These long-lived trees are also widely cultivated, for their majestic form, their brilliant, cinnamon brown bark, and their fanlike sprays of pungently fragrant foliage. The soft, finely textured wood, which does not splinter, is used in the manufacturing of most of the world's pencils; the scent of an incense cedar pencil can quickly summon memories of childhood classrooms. Even though plants can effectively stimulate memories with their unique scents (the tang of California sagebrush on a summer hike, the melancholy scent of decomposing sycamore leaves in winter, the perfume of a freshly cut lawn in spring), we have very few words for scents, usually defaulting to the phrase "it smells like…" In contrast, we have a vast number of terms for colors, often with several words referring to very similar colors (e.g., fuchsia, pink, fandango, cerise, magenta, ruby, etc.). This is possibly because humans have relatively low acuity to scents yet strong visual perception.

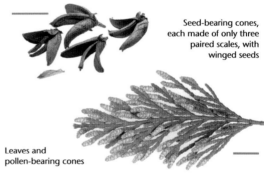

Seed-bearing cones, each made of only three paired scales, with winged seeds

Leaves and pollen-bearing cones

Some national parks have long waiting lists for camping reservations. When you have to wait a year to sleep next to a tree, something is wrong. —George Carlin

California's Most Fragrant Trees	
Incense Cedar	*Calocedrus decurrens*
Camphor Tree	*Cinnamomum camphora*
Lemon Scented Gum	*Corymbia citriodora*
Sweetshade	*Hymenosporum flavum*
Glossy Privet	*Ligustrum lucidum*
Sweetgum	*Liquidambar styraciflua*
Southern Magnolia	*Magnolia grandiflora*
Flowering Crab Apple	*Malus × floribunda*
Victorian Box	*Pittosporum undulatum*
California Bay Laurel	*Umbellularia californica*

Cedrus deodara

SEE-druss dee-uh-DOOR-uh

Kedros - Gr., cedar

deodara - Hindi, tree of the gods

Pinaceae

Himalayas

Needle-like, Alternate

Evergreen, 80 ft. or more

C. deodora leaves and female cone

T he four true cedars have a wide but discontinuous natural distribution from the Atlas Mountains of Morocco east to the Indian Himalayas. They are stately, evergreen conifers with aromatic wood and ten or more needle-like leaves grouped together on short shoots. Of the four cedars, the deodar cedar (*Cedrus deodara*) is planted most commonly in California, followed by the blue atlas cedar (*C. atlantica* var. *glauca*). The two species, which are closely related and occasionally hybridize, can be distinguished by the drooping branchlets of the deodar and the pale bluish color of the blue atlas. The other two species, cedar of Lebanon (*C. libani*) and Cyprus cedar (*C. brevifolia*) are rarely grown in California. Seed-bearing cones, rarely produced by trees younger than 40 years, look like small, erect footballs borne above the branches. The cones take two to three years to mature, at which point they break apart, releasing winged seeds. Deodar cedars are prized in California as specimen trees, lending a look of majesty to the gardens and parks where they are found. Resinous cedar wood is soft but durable and is used in carpentry and in furniture and cabinet making. More than 40 trees throughout the world bear the common name "cedar" but are not in the genus of the true cedars (*Cedrus).* For example, in California we have the incense cedar (*Calocedrus decurrens*), the western red cedar (*Thuja plicata*), and the Port Orford cedar (*Chamaecyparis lawsoniana*), among others.

The righteous shall grow like a cedar in Lebanon. —Psalms (92:12)

C. atlantica var. *glauca* leaves

C. deodara

Cupressus sempervirens   Italian Cypress

koo-PRESS-us sem-per-VYE-renz
Cupressus - L., for the Italian cypress
sempervirens - L., always green
Cupressaceae

 Mediterranean to Iran
Scale-like, Opposite
Evergreen, 60 ft.

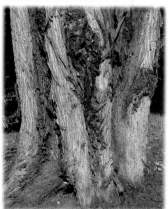

Trunk and bark of an older tree

The Italian cypress is the quintessential tree of Mediterranean landscapes, especially symbolic of Italy's Tuscan countryside, but it was introduced to Italy in ancient times; it is thought to have originated in Crete, Cyprus, Turkey, and Northern Iran. Outside of California, this species is most commonly known as Mediterranean cypress. It is often overplanted, but when used sparingly it can impart a look of stateliness to a garden. This long-lived conifer is easily recognized at a distance by its improbably slim, dark green crown, and up close by its egg-shaped seed cones and minute, scale-like leaves. Of all California's commonly grown trees, this species may have the smallest leaves; try to count the number on a small sprig and you will find thousands! The most columnar forms are propagated by cuttings and sold with various cultivar names, the most common of which is 'Stricta'. The Italian cypress has highly scented and very durable wood that was used for ornaments on Egyptian tombs, statues of gods in ancient Greece, and for building the doors of the Vatican's St. Peter's Basilica.

Stems, leaves, and female cones

Female cone and seeds

In trees, as in humans and animals, old age deserves respect and veneration. —Roger Underwood

Ginkgo biloba   Ginkgo or Maidenhair Tree

GING-koe bye-LOE-bah

Gin - Chinese, silver; *kyo* - Chinese, apricot

biloba - L., two-lobed

Ginkgoaceae

Southeastern China
Simple, Alternate
Deciduous, 40–80 ft.

Fleshy seeds

The ginkgo is unique among all plants as a lone survivor from the time of the dinosaurs. One of the world's oldest remaining species, its closest relatives went extinct millions of years ago. It can be readily identified by its fan-shaped leaves, an extract of which has been used for centuries to improve brain function. The ginkgo is popular in temperate cities all over the world because of its resilience to the tribulations of urban life, and for its golden fall leaf color. Most culti-vated trees are now grafted males because the overripe fleshy seeds of female trees have a foul stench, reminiscent of vomit and dog feces. Ginkgo may have become extinct in the wild sometime in the past few thousand years, only surviving in cultivation in China and later in Japan, where some trees are over one thousand years old. Some authorities believe that remnant wild stands still sur-vive in the mountains of south-eastern China.

Variation in
fall leaf color

Newly emerging leaves
and male cones

The passing of a great tree is never reported in the obituaries of our news-papers. This is a shame, for some of them have had unusual lives.
—Peter Del Tredici

Hesperocyparis macrocarpa Monterey Cypress

hess-pare-oe-sye-PARE-iss mack-roe-CAR-pah
Hesperos - Gr., western; *kyparissos* - Gr., cypress
macrocarpa - L., large fruit (cone)
Cupressaceae
Synonyms: *Cupressus macrocarpa, Callitropsis macrocarpa*

Monterey Peninsula, California
Scale-like, Opposite
Evergreen, 80 ft.

Although the Monterey cypress is widely cultivated in much of the western U.S., Europe, Australia, and New Zealand, in the wild it is one of the rarest trees in the world. It occurs naturally in only three foggy, windswept groves near Pebble Beach's Cypress Point and at Point Lobos, near the town of Carmel-by-the-Sea. This large and imposing tree can be recognized by its muscular trunk jacketed with thick, fibrous bark, its miniscule, overlapping, scale-like leaves, and its reddish brown, spherical cones. In cultivation a younger tree, invariably shaped like a pyramid, eventually becomes massive, picturesque, and irregular, with a horizontally tiered silhouette. Away from the coast, this species suffers, often fatally, from canker fungus infection. The old, unopened cones often visible on the branches attest to the fact that Monterey cypress is a canopy seed storer. These types of conifers bear cones with scales that are glued together with pitch and stay closed, keeping the seeds protected inside, until heated by a fire. This strategy allows for germination and regrowth only in the absence of competition from already established trees.

Plant trees. They give us two of the most crucial elements for our survival: oxygen and books. —A. Whitney Brown

Stems, leaves, and female cones

The Most Commonly Cultivated California Native Trees

California Buckeye	*Aesculus californica*
White Alder	*Alnus rhombifolia*
Incense Cedar	*Calocedrus decurrens*
Monterey Cypress	*Hesperocyparis macrocarpa*
Monterey Pine	*Pinus radiata*
California Sycamore	*Platanus racemosa*
Coast Live Oak	*Quercus agrifolia*
Valley Oak	*Quercus lobata*
Coast Redwood	*Sequoia sempervirens*
California Fan Palm	*Washingtonia filifera*

×*Hesperotropsis leylandii*   Leyland Cypress

hess-pare-oe-TRAWP-siss lee-LAND-ee-eye

Combined genera name - *Hesperocyparis* and *Callitropsis*

leylandii - Christopher John Leyland (1849–1926)

Cupressaceae

Hybrid

Simple, Opposite

Deciduous, 50–100 ft.

Synonyms: ×*Cuprocyparis leylandii, Cupressus* × *leylandii,* ×*Cupressocyparis leylandii*

The Leyland cypress is a ubiquitous and unremarkable, fast-growing conifer, often used as a dense screen or hedge. This sterile hybrid results from a cross between the Monterey cypress (*Hesperocyparis macrocarpa*, page 35) and the yellow cedar (*Callitropsis nootkatensis*), only in cultivation, as their native ranges are over four hundred miles apart. Taxonomists still debate the names of both closely related parents and their hybrid offspring, which have all been placed in different genera by various experts. For now, we'll call it the Leyland cypress, so named for a nineteenth-century horticulturalist who first developed this hybrid in England.

Branches with many scalelike, minute, opposite leaves

I've got to be by trees, otherwise I get claustrophobic.
—Liam Gallagher

Male cones

Female cones

Juniperus chinensis   Chinese or Hollywood Juniper

jew-NIH-perr-uss chih-NEN-siss
Juniperus - L., juniper
chinensis - L., from China
Cupressaceae

China, Mongolia, Japan
Scale-like, Opposite and Whorled
Evergreen, 20–40 ft.

Young cones

Individual,
minute, scale-
like leaves

Shoot with
adult leaves
and cones

"Tree" is not a distinct cate-
gory, like "dog" or "horse."
It is just a way of being a
plant. —Collin Tudge

uniperus is the largest genus in the cypress family (Cupressaceae), with its 50 or so species making up nearly half the family. Junipers are found throughout the northern hemisphere, from the equator all the way north to the extremes of the Arctic. The common juniper (*J. communis*), which is used to flavor gin, is the most widely distributed of all the world's conifers. Juniper cones, often referred to as berries, don't resemble typi-cal cones: they consist of a few fleshy scales coalesced into pulpy, bluish green, resinous globes. Unlike most woody cones, which typically open to release wind-borne seeds, juniper "berries" are often bird-dispersed, a likely reason for the cosmopolitan success of the genus. In Cali-fornia, which is home to four native junipers (including *J. communis*), many species are widely grown, but none more commonly than the Chinese juniper. This variable species has given rise to numerous cultivars, such as the ubiquitous Hollywood juniper (*J. chinensis* 'Torulosa' or 'Kaizuka'). The Chinese juniper can be recognized most easily by its mature form of twisted, awkward branches that rise waveringly yet emphatically skyward, each like a dark green flame. The Chinese juniper also makes two leaf types: opposite, scale-like, overlapping, adult leaves; and needle-like, spreading, juvenile leaves in whorls of three with a spiny, skin-piercing tip.

Juvenile
leaves

Adult
leaves

Metasequoia glyptostroboides   Dawn Redwood

meh-tuh-sih-KOY-yuh glip-toe-stroe-BOY-deez

Meta - Gr., beyond or after; *sequoia* - a related genus

glyptostroboides - resembling members of the
    genus *Glyptostrobus*

Cupressaceae

Central China
Needle-like,
Opposite
Deciduous, 90 ft.

F rom a group of trees that once spanned the entire Northern Hemisphere, there remain only three types of redwoods in the world: giant sequoia (*Sequoiadendron giganteum*), coast redwood (*Sequoia sempervirens*), and dawn redwood (*Metasequoia glyptostroboides*), and all are now relics with restricted natural ranges. Fossil records attest to various species of *Metasequoia* covering the Northern Hemisphere more than 65 million years ago, but as the planet cooled, the genus nearly went extinct, with the exception of this one remaining species. It too was presumed to be long-extinct and was known only from fossil records until surviving trees were discovered in 1941 in central China. Within a few years of this discovery, seeds were sent throughout the world, and now many have grown into large trees in California's parks and gardens. The dawn redwood is unlike the other redwoods in that its delicate foliage turns golden orange in the fall before dropping completely. The leaves are arranged in opposite pairs, each at right angles to the ones above and below. However, each pair is twisted at the base, bringing them all into the same plane.

Individual leaves

Spring shoot

To say that trees are immobile results from an anthropomorphism that impedes our seeing beyond our own time scale. It is as stupid as the history of aphids: In my memory, says the aphid, no one has ever seen a gardener die. Everyone knows that gardeners are immortal. —Francis Hallé

Deciduous
branchlets with
many pairs of
leaves in fall color

Seed cone

Pines make up the largest and most economically important genus of conifers. Their wood is widely used in construction and they are commonly grown as ornamentals. The nearly 100 species of pines are mostly in the Northern Hemisphere, with 20 native to California, yet only a small subset of these are commonly cultivated in the state's urban and suburban areas. Pines can be distinguished from other conifers by their needle-like leaves that are bundled together into fascicles surrounded by a membranous sheath, ranging from one to five needles per fascicle, depending on the species (see the image on page 43). Their winged seeds are formed in pairs in woody cones, on the upper face of the cone scales.

Drooping foliage of Jelecote Pine (*P. patula*)

Torrey Pine (*P. torreyana*)

Japanese Black Pine (*P. thunbergii*)

Aleppo Pine (*P. halepensis*)

Jelecote Pine (*P. patula*)

Mugo Pine (*P. mugo*)
All cones to this scale

Key to California's Commonly Cultivated Pines

1. Most bundles (fascicles) with 2 needles (occasionally with 3 needles)
   2. Mature plant a shrub or multi-branched small tree—Mugo Pine (*P. mugo*)
   2' Mature plant a large, single-stemmed tree
      3. Bark on old trunk breaking into large plates, some orangish in color; seed wing shorter than seed; tree crown rounded, umbrella-like—Italian Stone Pine (*P. pinea*)
      3' Bark on old trunk breaking into small or elongated plates, all brown or gray in color; seed wing longer than seed; tree shape varying
         4. Cones persisting for years (old branches with many cones)
            5. Needles mostly less than 3 in. long, cones recurved on stems—Aleppo Pine (*P. halepensis*)
            5' Needles mostly 3 in. long or more, cones forward-pointing on stems—Mondell Pine (*P. eldarica*)
         4' Cones falling at maturity (old cones not found on branches)
            6. Twigs often glaucous, buds chestnut brown; bark in upper part of tree orangish red, flaky—Japanese Red Pine (*P. densiflora*)
            6' Twigs not glaucous, buds conspicuously white; bark dark brown with deep longitudinal fissures—Japanese Black Pine (*P. thunbergii*)
1' Most bundles (fascicles) with more than 2 needles (rarely with only 2 needles)
   7. Needles 5 per bundle—Torrey Pine (*P. torreyana*)
   7' Needles 3 per bundle
      8. Foliage clearly drooping, cones less than 3 in. long—Jelecote Pine (*P. patula*)
      8' Foliage not drooping, cones greater than 3 in. long
         9. Needles mostly less than 6 in. long, cones asymmetrical—Monterey Pine (*P. radiata*)
         9' Needles 6 in. long or longer, cones symmetrical
            10. Cone scale with a blunt prickle, not sharp to the touch—Canary Island Pine (*P. canariensis*)
            10' Cone scale with a pointed prickle, sharp to the touch—Ponderosa Pine (*P. ponderosa*)

Winged seeds (*P. halepensis*)

Leaf cluster (fascicle) with a basal sheath

Pinus canariensis   Canary Island Pine

PYE-nus kuh-nair-ee-EN-sis

Pinus - L., pines

canariensis - L., of the Canary Islands

Pinaceae

Canary Islands
Needle-like, Alternate
Evergreen, 60–80 ft.

The Canary Island pine, the most widely cultivated pine in all of California, can be identified by its long graceful needles bundled in threes, its narrowly upright growth, its somewhat pendulous branches, and its rough, deeply fissured, reddish brown bark. The female cones are shiny brown with a blunt projection at the tip of each scale, and the smaller, ephemeral, male cones are responsible for the yellow clouds of allergenic pollen wafting through the air in late spring. This species has the ability to sprout new shoots from dormant buds under the bark (called epicormic sprouting), usually after some kind of damage like a fire. Some cultivated trees produce these bluish gray, waxy (glaucous) juvenile shoots even in the absence of any apparent injury; look for them on the lower trunk.

*Even if I knew that
tomorrow the world would
go to pieces, I would still
plant my apple tree.
—Martin Luther King Jr.*

Bark and epicormic juvenile sprout

Cluster of male cones

Female cone

Blunt cone
scale
prickle

Winged seed

Pinus halepensis   Aleppo Pine

PYE-nus hal-leh-PEN-siss

Pinus - L., pine

halepensis - L., from Aleppo, Syria

Pinaceae

Synonym: *Pinus halepensis* var. *brutia*

Mediterranean
Needle-like, Alternate
Evergreen, 40–80 ft.

The Aleppo pine is a drought-tolerant, two-needled pine from the Mediterranean with a large, irregular canopy and broad, spreading branches. A related species is also commonly grown in California, the Mondell or Afghan pine (*P. eldarica*). They're distinguished by the fact that the asymmetrical, persistent cones of the Aleppo pine are recurved backward on the branches, whereas the cones of the Mondell pine point outward. These resilient pines tolerate and even thrive in extreme temperatures, poor soils, drought, pollution, salt spray, and just about every other insult bestowed on neglected urban trees.

I love all trees, but I am in love with pines. —Aldo Leopold

P. halepensis male cones

P. halepensis recurved
female cones

Afghan Pine (*P. eldarica*) canopy with outward pointing cones

Gymnosperm: Conifer : **41**

Pinus pinea   Italian Stone Pine

PYE-nus PYE-nee-ah
Pinus - L., pines
pinea - L., pinecone
Pinaceae

Northern Mediterranean
Needle-like, Alternate
Evergreen, 40–80 ft.

Female cone with overlapping scales, each with two large, wingless seeds

The Italian stone pine is the world's best known source of pine nuts, which are actually seeds and not nuts (nuts are a type of fruit). A large portion of the world's culinary pine nuts come from wild and cultivated trees in Italy. In California, where the Italian stone pine is grown as a street and park tree, pine nuts can be gathered from freshly fallen cones. Pine nuts are a wonderfully tasty and nutritious food, a great source of protein, thiamin, phosphorous, iron, and, with an oil high in unsaturated fat, energy. The Italian stone pine can be recognized by its distinctive scaly orange-brown bark, needles bundled in pairs, and characteristically rounded, umbrella-shaped crown, often with multiple trunks.

Seeds with woody seed coats

In interactions with trees, Homo sapiens' *closest counterpart in the plant kingdom, humans express fundamental relationships with the nonhuman that help define who we are. The dividing line between humans and trees can become thin. —Kit Anderson*

Bark

Pinus radiata   Monterey Pine

PYE-nus ray-dee-AWE-tah

Pinus - L., pines; *radiata* - L., radiating

Pinaceae

Central California coast

Needle-like, Alternate

Evergreen, 80 ft.

The Monterey pine is rare in the wild, occurring only in three isolated stands along the California coast near Cambria, on the Monterey Peninsula, and along the coast north of Santa Cruz. In cultivation, this picturesque pine grows rapidly (up to 50 feet in ten years!) and is one of California's most popular large park trees. Rapid growth, a straight trunk with few lower branches, and quality wood have made the Monterey pine the most important plantation timber tree in the Southern Hemisphere. Selectively bred, genetically improved varieties are extensively planted (and have also become invasive weeds) in Australia, New Zealand, Chile, and South Africa. The largest cultivated forest in the world is a Monterey pine planta-

Lopsided female cone

tion called the Kaingaroa Forest on New Zealand's North Island. This species is extremely susceptible to pine pitch canker (*Fusarium circinatum*), a fungal disease from the southeastern U.S. first found in California in the 1980s, that has recently killed many wild and cultivated trees in the state. The Monterey pine can be recognized by its dark green needles that are bundled in threes (occasionally twos), its light brown, lopsided cones, and its dark gray bark.

The oaks and the pines, and their brethren of the wood, have seen so many suns rise and set, so many seasons come and go, and so many generations pass into silence, that we may well wonder what "the story of the trees" would be to us if they had tongues to tell it, or we ears fine enough to understand.

—Maud van Buren

A bundle of
needles (fascicle)
with 3 leaves

Sequoia sempervirens   Coast Redwood

sih-KOY-yuh sem-per-VYE-renz

Sequoia - Cherokee George Gist, also
    known as Sequoyah (1767–1843)

sempervirens - L., always green

Cupressaceae

California coast, Oregon coast

Needle-like, Alternate

Evergreen, 90–350 ft.

C alifornia's official state tree, the coast redwood, is among the world's most magnificent and revered organisms. They are surviving relics from a time long past when they were much more widely distributed. They have slowly retreated in their native range, surviving only in the specific climatic conditions of coastal California and Oregon, where they receive nearly one third of their water from condensed fog. They are capable of growing to great age, size, and height, with the most recently discovered tallest individuals stretching over 370 feet. These graceful giants are grown as large park, shade, and garden trees. Coast redwood lumber is straight-grained, clear, easily worked, and durable. Regrettably, ruthless logging of the nearly two million acres of virgin stands has left us with less than 5 percent of the original forests. At one time cheap and abundant (San Francisco was built twice with it), redwood lumber, especially from old trees, is now rare.

Leaves growing in a
plane on a slender stem

Female seed cones

Bluish green
cultivar with young
male cones

Spongy, cinnamon
brown, fibrous bark

It is foolish to let
a young redwood
grow next to a
house. Even in this
one lifetime, you
will have to choose.
That great calm
being or this clutter
of soup pots and
books—already the
first branch-tips
brush at the win-
dow. Softly, calmly,
immensity taps at
your life.
—Jane Hirshfield

Angiosperms

Acacia spp.

uh-KAY-shuh   *One way of measuring a tree is to fall from it. —Unknown*

Akakia - Gr., thorny plant in the bean family

Fabaceae

A s the world's largest tree genus, *Acacia* has over 1,200 species, found in warm, subtropical, and tropical regions of both hemispheres. Acacias are especially abundant in Australia, where the 900 or so species are the national floral emblem and generally referred to as "wattles." Acacias are the lonely, silhouetted trees of tropical African savannas, the large Australian shade trees, the solitary shrubs of Mexican and southwestern U.S. deserts, and the spiny, thicket-forming trees of Central America and South Africa. They are used worldwide for medicine, hardwood timber (especially blackwood, *A. melanoxylon*, and koa, *A. koa*), tannins, fuel, forage, and dyes, as well as water-soluble gums (gum arabic) used as thickening agents in processed foods and pharmaceuticals. Fast-growing acacias are regularly planted for land reclamation and restoration, and many species are cultivated ornamentally, including dozens in California. They are known for their aggressive reproduction and have become destructive weeds in many areas where they are widely cultivated, especially South Africa.

Knife Acacia (*A. cultriformis*)
leaves and flowers

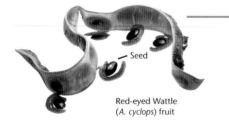

Red-eyed Wattle
(*A. cyclops*) fruit

Seed

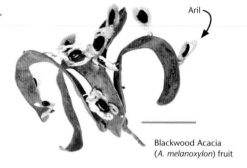

Aril

Blackwood Acacia
(*A. melanoxylon*) fruit

Hobo Trees

**Common Trees along California's
Roadways and Railroad Tracks**

Acacias (*A. retinodes, A. dealbata,*
    A. longifolia, A. melanoxylon)

Tree of Heaven (*Ailanthus altissima*)

Eucalypts *(Eucalyptus camaldulensis,*
    E. conferruminata, E. globulus)

Tea Tree (*Leptospermum laevigatum*)

Glossy Privet (*Ligustrum lucidum*)

Myoporum (*Myoporum laetum*)

Victorian Box (*Pittosporum undulatum*)

Coast Live Oak (*Quercus agrifolia*)

Coast Redwood (*Sequoia sempervirens*)

Peruvian Pepper (*Schinus molle*)

Silver Wattle (*A. dealbata*)

☿ ♂ Tropics, subtropics worldwide
Bipinnate and Phyllodes, Alternate
Evergreen, 10–100 ft.

All acacias make feathery, bipinnately compound leaves, but in some species, phyllodes replace the compound leaves shortly after germination. A phyllode is a leaflike structure derived from the flattening of the leaf stalk (petiole) of a compound leaf. These photosynthetic structures, which have a reduced surface area compared to true bipinnately compound acacia leaves, are found mostly in Australian species and are thought to have evolved in drought-prone areas. Around the base of a blackwood acacia (*A. melanoxylon*), one can often find seedlings that are in the process of transitioning from feathery, compound leaves to flattened, leathery phyllodes. Most acacias are short lived (20 to 30 years) and bloom profusely when young. Their minute flowers are cream-colored to bright yellow and arranged in dense globular or cylindrical clusters. Acacia fruits (legumes) are filled with flat beans that in some species are surrounded by a fleshy, colorful tissue (aril) that attracts ants or birds and thus enhances seed dispersal.

*Whether they belong to an Egyptian temple, the Parthenon, or
Amiens cathedrals, the columns that enclose the presence of gods
are imitations of the trunks of ancient trees. —Steven Marx*

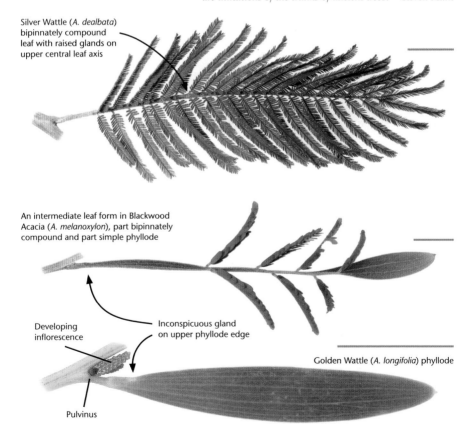

Silver Wattle (*A. dealbata*)
bipinnately compound
leaf with raised glands on
upper central leaf axis

An intermediate leaf form in Blackwood
Acacia (*A. melanoxylon*), part bipinnately
compound and part simple phyllode

Developing
inflorescence

Inconspicuous gland
on upper phyllode edge

Golden Wattle (*A. longifolia*) phyllode

Pulvinus

Acacia spp.

*Hugging trees has a calming effect on me. I'm talking about enormous
trees that will be there when we are all dead and gone. —Gerry Adams*

Star Acacia (*A. verticillata*)

Shoestring Acacia (*A. stenophylla*)

Golden Wreath
Wattle (*A. saligna*)
flowers

Black Wattle (*A. mearnsii*),
arrow indicates leaf gland

Golden Wattle (*A. longifolia*) flowers

Golden Wreath Wattle
(*A. saligna*) leaf blade
showing conspicuous
gland

Leaf
pulvinus

Green Wattle
(*A. decurrens*) hairless
stems with wings

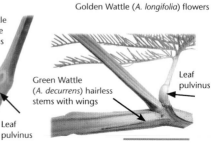

Leaf
pulvinus

Leaf
pulvinus

Tropics, subtropics worldwide
Bipinnate and Phyllodes, Alternate
Evergreen, 10–100 ft.

Key to California's Commonly Cultivated Acacias

1. Leaves bipinnately compound
   2. Leaves strongly glaucous (covered with wax that can be rubbed off), occasionally purple, with 3 to 5 pairs of primary leaflets—Bailey Acacia *(A. baileyana)*
   2′ Leaves not distinctly waxy, usually green, sometimes silvery or bluish gray hairy, with 6 or more pairs of primary leaflets
      3. Branchlets hairless, with pronounced winglike ridges, especially at nodes; primary leaflets not overlapping—Green Wattle *(A. decurrens)*
      3′ Branchlets with downy hair, angled, but not with pronounced wings; primary leaflets overlapping
         4. Lower surface of leaflets more or less the same color as the upper surface, both sparsely hairy; raised glands at attachment point of primary leaflets; fruit hairless, silver-blue; flowers golden—Silver Wattle *(A. dealbata)*
         4′ Lower surface of the leaflets distinctly paler (with silvery hairs) than the upper surface (hairless, green); raised glands between attachment points of primary leaflets; fruit hairy, dark brown; flowers cream-colored—Black Wattle *(A. mearnsii)*
1′ Leaves simple (phyllodes)
   5. Phyllodes needle-like, linear, <⅛ in. wide, spiny, whorled in threes—Star Acacia *(A. verticillata)*
   5′ Phyllodes blade-like >¼ in. wide, alternate
      6. Phyllodes <1.5 in. long, triangle-shaped—Knife Acacia *(A. cultriformis)*
      6′ Phyllodes >2 in. long
         7. Tree weeping, with pendent, hanging branches
            8. Phyllodes covered with silvery gray down—Weeping Acacia *(A. pendula)*
            8′ Phyllodes hairless—Shoestring Acacia *(A. stenophylla)*
         7′ Tree or shrub with upright branches
            9. Phyllodes with one conspicuous central longitudinal nerve or vein
               10. Stems round, >10 flowers per flower cluster—Golden Wattle *(A. pycnantha)*
               10′ Stems ridged, <10 flowers per flower cluster
                  11. Leaf base with obvious gland (⅛ in. wide) near pulvinus—Golden Wreath Wattle *(A. saligna)*
                  11′ Leaf base with inconspicuous gland—Everblooming Acacia *(A. retinodes)*
            9′ Phyllodes with two or more conspicuous longitudinal veins, or obscure veins and nerves
               12. Plant a large, single-trunked tree; phyllodes dark green; flowers cream-colored—Blackwood Acacia *(A. melanoxylon)*
               12′ Plant a densely branched shrub; phyllodes light green; flowers bright yellow
                  13. Flowers clustered in cylindrical spikes—Golden Wattle *(A. longifolia)*
                  13′ Flowers clustered in spherical heads
                     14. Phyllodes generally ½ in. or more wide, bluish green; plant often a ground cover—Bank Catclaw *(A. redolens)*
                     14′ Phyllodes generally ¼ in. or less wide, olive green; plant an erect shrub—Red-Eyed Wattle *(A. cyclops)*

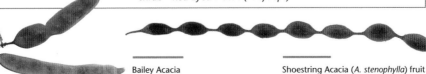

Bailey Acacia
(*A. baileyana*) fruit

Shoestring Acacia (*A. stenophylla*) fruit

Acacia baileyana   Bailey Acacia

uh-KAY-shuh bay-lee-AWE-nuh

Akakia - Gr., thorny plant in the bean family

baileyana - Frederick Manson Bailey (1827–1915)

Fabaceae

Southeastern Australia

Bipinnate, Alternate

Evergreen, 20–30 ft.

Flower clusters

The Bailey acacia, which is called Cootamundra wattle in Australia, is California's second most commonly planted acacia. Its profuse, vibrant, golden midwinter blooms are formed above a background of grayish blue or grayish lavender ('Purpurea' cultivar) foliage. This species is popular as a shapely, midsized street tree in cooler coastal cities. It can be recognized by its smooth, blue-tinted, gray trunk and waxy, bipinnately compound leaves, each with several spherical nectar glands along the upper side of the leaf axis. In the wild, these glands exude sugar droplets from nectaries within, an adaptation that attracts stinging ants that protect the trees from herbivores.

A leaf showing
the glands on the
upper surface of
the leaf stalk

For all its mass, a tree is a remarkably delicate thing. All of its internal life exists within three paper-thin layers of tissue—the phloem, xylem, and cambium—just beneath the bark, which together form a moist sleeve around the dead heartwood. However tall it grows, a tree is just a few pounds of living cells thinly spread between roots and leaves. These three diligent layers of cells perform all the intricate science and engineering needed to keep a tree alive, and the efficiency with which they do it is one of the wonders of life. —Bill Bryson

Acacia melanoxylon   Blackwood Acacia

uh-KAY-shuh mel-ah-NOX-sih-lawn

Akakia - Gr., thorny plant in the bean family

melas - Gr., black; *xylon* - Gr., wood

Fabaceae

Eastern Australia

Simple (phyllode), Alternate

Evergreen, 60+ ft.

The blackwood acacia is the most commonly grown acacia in California. It can be found throughout the western portion of the state as a street and shade tree, and occasionally as a weed along roadways on the periphery of urban areas. This is one of the world's largest and longest-lived acacias, reaching nearly one hundred feet in height in its native southeastern Australia, where it is logged for valuable hardwood used in cabinetry and furniture making. Its unremarkable, cream-colored, spring flowers are followed by prodigious numbers of sickle-shaped legumes that split to reveal shiny black seeds. A thickened stalk that resembles an umbilical cord, called the funiculus, coils twice around each seed, tethering it to the pod as it matures. When the seeds ripen and the fruits open, birds and ants carry the funiculi away, along with the seeds attached to them; when they eat the funiculi, the seeds are discarded into new environs.

Flowers

The cultivation of trees is the cultivation of the good, the beautiful, and the ennobling in man. —J. Sterling Morton

Acacia melanoxylon

Bunya Bunya
(*Araucaria bidwillii*)

Acer spp.

AYE-ser

Acer - L., maples

Sapindaceae

The 114 species of maples are found almost entirely in the north temperates, with a great many species in China, one in northern Africa, and a number in high-elevation areas of the Asian tropics. They range greatly in size from dwarf shrubs to tall trees, and like the oaks, some are evergreen and others are deciduous. Maples are easily recognized by their fruits: a pair of winged segments, each containing a seed. These wind-dispersed fruits (called "samaras" or "keys") are perfectly weighted to spin helicopter-like through the air when dislodged or detached from the tree. Pick a fruit off the next maple you see and throw it skyward and you will be surprised by how well it flies. It helps to shout, "Be free!" as you throw it. Maple leaves are opposite on the stem and are usually palmately lobed. Their flowers are relatively inconspicuous, generally wind-pollinated, and emerge as hanging, greenish yellow clusters as new spring leaves unfold.

Maples are celebrated throughout the northeastern U.S. and Canada for their brilliant fall color. One of the species found there, the sugar maple (*Acer saccharum*) is the source of the light-colored, hardwood timber used commonly in furniture, flooring (particularly gymnasiums), and instrument making. The sap of this species is concentrated to form the delicious maple syrup that is capable of elevating morning pancakes from mediocrity to the sublime. In California, which is home to five native species, most maples are grown as ornamentals on streets, in parks, and in gardens in the cooler, wetter, northwestern part of the state. In general, maples don't grow well in the dry heat of Southern California.

Japanese Maple
(*A. palmatum*) leaves
and flowers

Norway Maple
(*A. platanoides*)
fruit

Hedge Maple
(*A. campestre*)
leaves and fruit

Japanese Maple
(*A. palmatum*)
fall foliage

Silver Maple
(*A. saccharinum*)
trunk

North Temperates, Tropical Asia
Simple or Compound, Opposite
Deciduous and Evergreen, 20–90 ft.

The most commonly grown species throughout the entire state are Japanese maple (*A. palmatum*) and red maple (*A. rubrum*). Japanese maples, which rarely top out over 20 feet, are delicate garden trees, with dozens of cultivars. Silver maples are common even in Southern California and have gray bark that peels in long strips and bright, golden fall color. The showy red maple has brilliant, crimson fall color, especially in frostier parts of California.

Key to California's Commonly Cultivated Maples

1. Leaves compound (with three or more leaflets)—Box Elder (*A. negundo*) **4**
1′ Leaves simple
  2. Leaf undersides hairy, white, grayish white, or bluish gray
    3. Most leaves 1 to 3 in. wide
      4. Leaves usually not lobed, sometimes 3-lobed in
        young shoots—Smooth Leaf Maple (*A. oblongum*) **12**
      4′ Leaves always 3-lobed
        5. Leaf lateral lobes nearly same size as middle lobe—
          Trident Maple (*A. buergerianum*) **11**
        5′ Leaf lateral lobes usually smaller than middle lobe,
          leaves occasionally unlobed—Evergreen Maple (*A. paxii*) **3**

    3′ Most leaves >3 in. wide
      6. Middle lobe of most leaves wider near the middle than at the base,
        leaf underside bright white—Silver Maple (*A. saccharinum*) **5**
      6′ Middle lobe of most leaves widest at the base,
        leaf underside light gray, grayish white, or bluish gray
        7. Lobe margins smooth or wavy, or with
          rounded teeth—Sugar Maple (*A. saccharum*) **7**
        7′ Lobe margins irregularly or sharply toothed
          8. Leaves 2 to 4 in. wide, lobes sharply toothed,
            pointed at tip—Red Maple (*A. rubrum*) **8**
          8′ Leaves 3 to 7 in. wide, lobes with round teeth,
            rounded at tip—Sycamore Maple (*A. pseudoplatanus*) **6**
  2′ Leaf undersides hairless, green (although usually paler than the upper side), not white
    9. Leaves with 5 to 11 lobes, tree often small and
      multistemmed—Japanese Maple (*A. palmatum*) **9**
    9′ Leaves with 3 to 7 lobes, tree larger, often single stemmed
      10. Most leaves <3 in. wide—Hedge Maple (*A. campestre*) **10**
      10′ Most leaves >3 in. wide
        11. Most leaves 6 to 12 in. wide, lobes cut deeper than
          the middle of the leaf—Big Leaf Maple (*A. macrophyllum*) **1**
        11′ Most leaves 3.5 to 8 in. wide, lobes not deeper than the middle of the leaf
          12. Sap in petiole white; tips of leaf points reduced to a fine hair—Norway
            Maple (*A. platanoides*) **2**
          12′ Sap in petiole clear; tips of leaf points not reduced to a hair—Sugar
            Maple (*A. saccharum*) **7**

Aesculus × *carnea*

ESS-kew-lus kar-NEE-ah
Aesculus - L., oak with edible acorn
carnea - L., flesh-colored
Sapindaceae

Hybrid
Palmate, Opposite
Deciduous, 40 ft.

California buckeye
(*A. californica*) fruits and seeds

Red horse
chestnut leaf

Members of the genus *Aesculus* are called "horse chestnuts" in Europe or "buckeyes" in the U.S. due to the seed's resemblance to a deer's eyeball. Trees with palmately compound leaves, attached two per node (oppositely), are rare in the temperates, making horse chestnuts and buckeyes easily recognized. Red horse chestnut is a fertile hybrid between the North American red buckeye (*A. pavia*) and one of Europe's most famous street and park trees, the European horse chestnut (*A. hippocastanum*). It can be recognized while leafless in midwinter by its unusually large, slightly sticky terminal buds. These buds burst open in early spring, forming erect spires of crimson, rose red, and deep pink flowers. The native California buckeye (*A. californica*), which is also occasionally cultivated as a park and garden tree, can be distinguished by its white- and pink-tinged flowers and stalked leaflets. All parts of horse chestnuts and buckeyes contain the poisonous glycoside aesculin, and human fatalities occasionally occur, mostly from children eating the seeds.

Everyone who enjoys thinks that the principal thing to the tree is the fruit, but in point of fact the principal thing to it is the seed. Herein lies the difference between them that create and them that enjoy. —Friedrich Nietzsche

Terminal bud

Deciduous winter stem

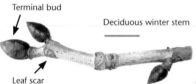

Leaf scar

European horse chestnut (*A. hippocastanum*)

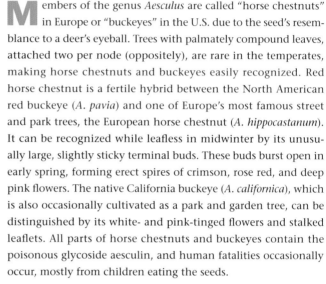

A. × *carnea* flowers

Agonis flexuosa   Willow Peppermint

ay-GOE-niss flex-yoo-OH-sah

Agon - Gr., an assembly or collection

flexuosa - L., bending, zigzagged

Myrtaceae

Southwestern Australia

Simple, Alternate

Evergreen, 40 ft.

The willow peppermint is an attractive, small tree with a graceful weeping form, ideal for coastal sites and under power lines. It was brought into cultivation in California in the early 1870s from the limestone heathlands and coastal dunes of the Mediterranean climate region of southwestern Australia, near Perth. Although it evolved in a relatively dry climate on sandy soils, given good drainage it will grow well in varying moisture conditions on most soil types, in lawns, parkways, and gardens. The willow peppermint can be recognized by its twigs, which have a distinctly zigzagged form, and its linear, olive green, willow-like leaves that smell subtly of peppermint when crushed. Holding a leaf to the light reveals hundreds of clear dots (oil glands). In spring, five-petaled, fragrant, white flowers emerge, followed by spherical clusters of woody, cup-shaped fruits. The bark on mature trees is deeply furrowed, fibrous, and reddish brown. The willow peppermint is a member of the widely cultivated myrtle family (Myrtaceae), which includes the eucalypts, bottlebrushes, tee-trees, and melaleucas, but there are no native members of this family in California.

A stricken tree, a living thing,
so beautiful, so dignified,
so admirable in its potential
longevity, is, next to man,
perhaps the most touching
of wounded objects.
—*Edna Ferber*

Branchlet with typical
zigzagged form

Bark

Ailanthus altissima   Tree of Heaven

aye-LAN-thus al-TISS-ih-mah
Ailanto - Moluccan, sky tree
altissima - L., very tall
Simaroubaceae

China
Pinnate, Alternate
Deciduous, 50 ft.

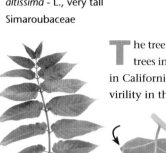

Leaf

Leaflet underside
with two glands

The tree of heaven is considered by some to be one of the weediest trees in the world. Although it hasn't been intentionally planted in California for years, it still grows and reproduces with incredible virility in the state's forgotten landscapes: in back alleys and vacant lots, along roads and railroad tracks, and at abandoned old homesteads, especially near previous Chinese laborer camps in the Sierra Nevada foothills. This species is famous for its ability to grow and thrive in the worst situations, hence the pejorative common name "ghetto palm." It was the titular tree and central metaphor for thriving in difficult conditions in Betty Smith's famous 1943 novel *A Tree Grows in Brooklyn*. The short-lived, fast-growing tree of heaven has foul-smelling flowers and a suckering habit whereby it produces copious sprouts directly from shallow roots. It manufactures chemicals that kill other plants and is nearly impossible to kill. What's not to love?

Winged fruits

There's a tree that grows in Brooklyn. Some people call it the Tree of Heaven. No matter where its seed falls, it makes a tree which struggles to reach the sky. It grows in boarded up lots and out of neglected rubbish heaps. It is the only tree that grows out of cement. It grows lushly... without sun, water, and seemingly earth. It would be considered beautiful except that there are too many of it.
—*Betty Smith*, A Tree Grows in Brooklyn

Albizia julibrissin   Silk Tree

al-BEE-zee-ah joo-lih-BRISS-in
Albizia - Filippo del Albizzi
gul - Persian, flower; *abrisham* - Persian, silk
Fabaceae

Iran to China, Korea
Bipinnate, Alternate
Deciduous, 40 ft.

With 19,500 species, the legume family (Fabaceae) is the third-largest plant family, surpassed only by the orchid family (22,500 species) and the sunflower family (23,600 species). Legumes make three types of flowers that define three subfamilies. The Caesalpinioideae (*Bauhinia, Cercis, Ceratonia, Cassia*) are usually tropical trees with large-petaled, bilaterally symmetrical flowers and pinnate leaves. The Papilionoideae/Faboideae (*Robinia, Gleditsia, Erythrina, Tipuana*) are widely distributed trees, shrubs, and herbs, with the typical bilaterally symmetrical, pea-shaped flowers. The silk tree is a member of the third subfamily, Mimosoideae (*Acacia, Albizia*), which are usually tropical trees with bipinnate leaves and radially symmetrical flowers with tiny petals and numerous, prominent, often colorful stamens. For all its tropical appeal, (broad, dome-shaped crown, feathery, finely textured leaves, and pink, puffball flowers), the silk tree is surprisingly hardy and grows well in California's colder inland areas. Another characteristic common among many legumes is a swollen structure at the base of leaves and leaflets called a pulvinus (see photo on page 47). Leaves rise, fall, and orient themselves to the sun when the pulvinus swells and deflates. Viewing a silk tree at night will reveal a "sleeping tree" with folded leaflets on drooping leaves.

Alone with myself, the trees bend to caress me, the shade hugs my heart. —Candy Polgar

Fruits and seeds

Angiosperm: Eudicot: Fabales : **57**

Alnus spp.

ALL-nuss
Alnus - L., alder
Betulaceae

North Temperates
Simple, Alternate
Deciduous, 40–75 ft.

Tiny winged fruits

A. cordata

A. rhombifolia

O f the 30 or so species of alders, three are commonly cultivated in California: white alder (*Alnus rhombifolia*) and red alder (*A. rubra*), both California natives, and the Italian alder (*A. cordata*), native to Italy and Corsica. A year in the life of a typical alder would begin with a deciduous tree in early winter. As spring approaches, just before its new leaves emerge, it makes thousands of pendulous clusters of tiny, pollen-forming flowers and upright clusters of minute, female, egg-bearing flowers. With the long days of summer advancing, the tree reaches its full dark green, leafy glory and the female flowers mature within conelike structures made of many minute, woody scales. As these cones ripen in late summer and fall, each releases hundreds of tiny, winged fruits; the cones remain attached to the branches while the tree once again loses its leaves as winter draws near. Alders are related to birches (*Betula* spp.) but can easily be distinguished from them. The conelike structures of birches disintegrate during or after the time when they are releasing the winged fruits. Also, alders get nourishment from a symbiotic relationship with bacteria that inhabit their roots, fixing atmospheric nitrogen into usable fertilizer.

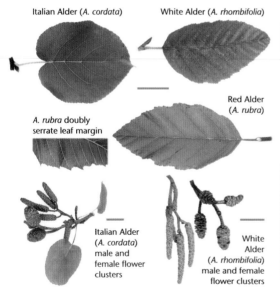

Italian Alder (*A. cordata*)

White Alder (*A. rhombifolia*)

A. rubra doubly serrate leaf margin

Red Alder (*A. rubra*)

Italian Alder (*A. cordata*) male and female flower clusters

White Alder (*A. rhombifolia*) male and female flower clusters

A. rhombifolia

What is homo but arbor inversa?
—John Evelyn

Key to Commonly Cultivated Alders
1. Leaves cordate (heart-shaped)—Italian Alder (*A. cordata*)
1' Leaves ovoid (egg-shaped) or rhomboid (diamond-shaped)
2. Leaves coarsely toothed, tightly rolled under at margins—Red Alder (*A. rubra*)
2' Leaves finely toothed, flat at margins—White Alder (*A. rhombifolia*)

arr-BYEW-tuss muh-REE-nuh

Arbutus - L., strawberry tree

Ericaceae

Hybrid

Simple, Alternate

Evergreen, 40 ft.

The strawberry tree is an elegant, well-behaved, medium-sized tree that has become popular in California's gardens, parks, and street wells. It is a hybrid with unknown parents (*Arbutus andrachne, A. canariensis,* and *A. unedo,* all from Europe, are possible parents), introduced into the nursery trade by the Saratoga Horticultural Foundation in 1984. The hybrid is named for the San Francisco Marina district, the location of a nursery that may have originally acquired the tree from Europe during the 1915 Panama-Pacific International Exposition. The strawberry tree can be recognized by its bark, which annually exfoliates to expose strikingly beautiful, smooth, reddish brown or cinnamon-colored new bark. Pendent clusters of rose pink, lantern-shaped flowers are followed by orange to red, warty, spherical fruits that taste like mealy apricots full of large-grained sand; not at all like strawberries. *Unedo,* the specific epithet of one of the possible parents, means "I eat (only) one."

Flowers

Developing fruits

Leaf

We still have this to learn: the inalienable otherness of each, human and non-human, which may seem the prison of each, but is at heart, in the deepest of those countless million metaphorical trees for which we cannot see the wood, both the justification and the redemption. —John Fowles

Bauhinia spp.

bauw-HINN-ee-ah
Bauhinia - Johann Bauhin (1541–1613)
and Gaspard Bauhin (1560–1624)
Fabaceae

Asian Tropics
Simple, Alternate
Evergreen, 20–30 ft.

Leaf

B. variegata open fruit

The orchid tree's dazzling flowers, white to rose pink to maroon, are among California's showiest. They superficially resemble orchids, but these trees are actually in the legume family. Although the genus has over 200 species, only three are commonly grown in California: the purple orchid tree (*Bauhinia variegata*), the white orchid tree (*B. forficata*), and the Hong Kong orchid tree (*B. × blakeana*); the last is completely sterile, never making seeds, and is propagated only by grafting. Carl Linnaeus named the genus *Bauhinia* after two sixteenth-century botanists, the Swiss brothers Johann and Gaspard Bauhin, using the twin leaf lobes, a characteristic of the genus, to symbolize their lifelong professional collaboration. Orchid trees generally don't survive freezing temperatures—a reason they are common in Southern California and rare in Northern California.

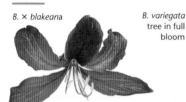

B. × blakeana

B. variegata tree in full bloom

B. galpinii

B. variegata

Sing and dance, make faces and give flower bouquets, trying to be loved. You ever notice that trees do everything to get attention we do, except walk?
— Alice Walker,
The Color Purple

California's Showiest Trees

Orchid Trees (*Bauhinia* spp.)
Floss Silk Trees (*Ceiba speciosa*)
Red Flowering Gum (*Corymbia ficifolia*)
Coral Trees (*Erythrina* spp.)
Trumpet Trees (*Handroanthus* spp.)
Sweetshade (*Hymenosporum flavum*)
Jacaranda (*Jacaranda mimosifolia*)
Crape Myrtles (*Lagerstroemia* spp.)
Saucer Magnolia (*Magnolia × soulangeana*)
Palo Verdes (*Parkinsonia* spp.)
Flowering Cherries (*Prunus* spp.)
African Tulip Tree (*Spathodea campanulata*)

Betula pendula   White Birch

BEH-tyoo-lah PEN-dyoo-lah

Betula - L., birch

pendula - L., hanging

Betulaceae

Northern Europe, Asia Minor

Simple, Alternate

Deciduous, 30–40 ft.

This graceful Northern European tree is grown throughout the western U.S. and California and is the most widely grown of the 35 species of birches. Their delicate appearance is misleading; birches endure severe winter conditions in their native habitat and can grow farther north than any other nonconiferous trees. In Northern Europe, white birch has been an important economic tree, used for furniture wood, leather tanning, famine food, and medicinally, to treat skin conditions. All birches are deciduous and wind-pollinated, with minute flowers borne in conelike structures that break apart when mature, releasing many tiny winged fruits. The white birch can be recognized by its weeping branches adorned with glossy, diamond-shaped, double-sawtoothed leaves and its distinctive bark. The bark is initially golden brown on the smaller twigs; it then becomes smooth and bright white on the larger branches and upper trunk, and it finally cracks to form black fissures on the oldest parts of the trunk. The numerous cultivated varieties of white birch include some with burgundy leaves and others with deeply divided leaves.

Leaves and immature catkins

Mature female catkin and winged fruits

I'd like to go by climbing a birch tree, and climb black branches up a snow-white trunk toward heaven, till the tree could bear no more, but dipped its top and set me down again. That would be good both going and coming back. One could do worse than be a swinger of birches. —Robert Frost

Brachychiton spp.

brak-ee-KYE-tun
Brachys - Gr., short
chiton - Gr., a tunic
Malvaceae

B. *acerifolius* seeds

B. *rupestris* flowers

O f the 31 species in the genus *Brachychiton*, which are mostly from eastern Australia, four are regularly found in California: Kurrajong (*B. populneus*), Illawarra flame tree (*B. acerifolius*), lacebark (*B. discolor*), and Queensland bottle tree (*B. rupestris*), listed in order of prevalence. There are also several named hybrids between species. The swollen, greenish brown trunks of this genus bulge in the middle like Greek columns. Their leaves vary greatly, from diamond-shaped to deeply palmately lobed, and they tend to bloom erratically and only on a portion of the canopy. Most have elegant, bell-shaped flowers without petals but with colorful sepals. The heavy, woody, canoe-shaped fruits clap together in the wind like castanets. The roasted seeds are a traditional food of Indigenous Australian peoples, and fiber from the bark can be used to make baskets, nets, fishing line, and ropes. In California, *Brachychiton* grows best in milder climates, in well-drained soil and full sun.

B. *discolor* flower

For in the true nature of things, if we rightly consider, every green tree is far more glorious than if it were made of gold and silver. —Martin Luther

B. *acerifolius* trunk and canopy

B. *acerifolius* flower

Eastern Australia
Simple, Alternate
Evergreen or Partially Deciduous, 40–50 ft.

If there were but one erect and solid standing tree in the woods, all creatures would go to rub themselves against it, and make sure of their footing. —Henry David Thoreau

Fruit, *B. discolor* (top)

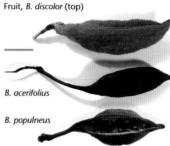

B. acerifolius

B. rupestris

B. populneus

B. populneus flowers

Typical leaf shapes and intraspecies variation

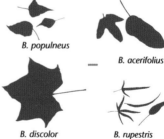

B. populneus

B. acerifolius

B. discolor

B. rupestris

Key to Commonly Cultivated *Brachychiton*

1. Leaves distinctly lighter and hairy on the underside—Lacebark *(B. discolor)*
1' Leaves mostly the same color on both sides, hairless
    2. Most leaves >4 in. wide; flowers scarlet red—Illawarra flame tree *(B. acerifolius)*
    2' Most leaves <3.5 in. wide; flowers white or cream-colored, and spotted
        3. Leaves on young trees entire or shallowly lobed; flowers white, hairless inside; mature trunk moderately swollen—Kurrajong *(B. populneus)*
        3' Leaves on young trees deeply lobed; flowers yellow or cream-colored, hairy inside; mature trunk conspicuously swollen in the middle—Queensland bottle tree *(B. rupestris)*

Callistemon spp.

kah-LISS-teh-mon

Kallos - Gr., beautiful; *stemon* - Gr., stamen

Myrtaceae

Synonym: *Melaleuca* in part

Callistemon is a genus of about 50 species entirely native to Australia, but widely culti- vated in many other regions. In California, bottlebrushes bloom irregularly throughout the year and then prodigiously in spring and summer. The flowers are filled with sugary nectar that attracts hum- mingbirds, other nectar feeders, and even the occasional squirrel. The flowers are followed by beadlike, woody cap- sules that can remain on branches for years and eventu- ally be consumed by the expanding bark.

Fruits

I can take something from a tree without killing it, like a bouquet of flowers or a basket of fruit. No animal would ever allow such dissection. —Yildiz Aumeeruddy

C. viminalis

C. citrinus

Australia
Simple, Alternate
Evergreen, 20–40 ft.

Melaleuca is closely related to *Callistemon* but differs in having stamens that are fused in five bundles. Some authorities now treat all *Callistemon* species as *Melaleuca*, though this taxonomy is still controversial. The two most commonly cultivated bottlebrushes in California, weeping bottlebrush (*C. viminalis*) and lemon bottlebrush (*C. citrinus*), have several distinguishing characteristics. The weeping bottlebrush is a single-stemmed tree with billowy, arching branches and crimson stamens connected in a ring. The smaller lemon bottlebrush is an erect, large shrub or multistemmed small tree (although it is often forcefully pruned into a single-stemmed tree) with leaves that are subtly lemon-scented when crushed, and stamens not fused in a ring at the base.

Basally fused stamens
of *C. viminalis*

C. citrinus
flowering shoot

C. viminalis

One of the many pink
cultivars of *C. citrinus*

Cassia leptophylla

Gold Medallion Tree

KASS-ee-ah lep-toe-FILL-ah

Kassia - Gr. name for biblical kasian plant

leptos - Gr., thin; *phyllon* - Gr., leaf

Fabaceae

Southeastern Brazil
Pinnate, Alternate
Evergreen, 20–30 ft.

With all its lovely features, one has to wonder why the gold medallion tree isn't grown more frequently. It was first planted at the Los Angeles Arboretum in 1958 and has become steadily more popular in California since. It is a fast-growing, relatively hardy tree with a tropical appeal and can be recognized by its rounded crown of glossy green, pinnate leaves that lack a terminal leaflet. The long-lasting, fragrant, deep-yellow flowers are produced intermittently throughout the year and extensively in the summer. These flowers develop into one to two-foot-long pods that are square in cross section.

Only when the last tree has died and the last river been poisoned and the last fish been caught will we realize we cannot eat money. —Cree Indian Proverb

——— Fruit

Casuarina spp.   Horsetail Trees

kazh-yoor-EYE-nah

Casuarina - L., drooping branchlets
   resemble cassowary feathers

Casuarinaceae

Eastern Australia
Scale-like, Whorled
Evergreen, 40–60 ft.

The horsetail trees (also known as "sheoaks") are native to watercourses in eastern Australia and are greatly valued for their tolerance of salt spray, high winds, and severe soil conditions. *Casuarina* species hybridize readily, and distinguishing them can be difficult, often requiring a magnifying lens. In general, *Casuarina* can be recognized by its slender, green branchlets that resemble pine needles and have tiny, scale-like leaves whorled around light brown joints. Male trees make minute, pollen-producing flowers arranged in spikes at the branch tips, whereas female trees make flowers in woody, conelike structures. The grooved, jointed branchlets superficially resemble the stems of horsetail plants (*Equisetum* spp.) and pine needles. In California, these rugged and resilient trees are grown as street and park trees and have been replacing blue gum (*Eucalyptus globulus*) as the agricultural windbreak tree of choice.

C. cunninghamiana flowers

*Solitary trees,
if they grow at
all, grow strong.*
—Winston Churchill

C. cunninghamiana
stems with minute,
whorled leaves

What *Casuarina* Is That?

Most of the sheoaks in California are *C. cunninghamiana* or the less common drooping sheoak (*Allocasuarina verticillata*: synonyms are *Casuarina verticillata* and *C. stricta*). If the branchlets are drooping and the distance between nodes is greater than ½ inch, the tree is likely *A. verticillata*. If the branchlets are erect and the distance between nodes is less than ½ inch, the tree is likely *C. cunninghamiana*. The name *Casuarina equisetifolia* (a species rarely found in California) is commonly and incorrectly used in reference to *C. cunninghamiana*.

Catalpa spp.

kuh-TAL-pah

Catawba - Native American tribe

Bignoniaceae

Midwestern United States

Simple, Opposite

Deciduous, 40–80 ft.

atalpas reached the peak of their popularity in California during the Victorian era, at the end of the nineteenth century, and have since been far less commonly planted. Their large, heart-shaped leaves that turn golden in the fall can create a lush, almost tropical appearance, even in the coldest parts of the state. Even though trees cannot run away to avoid being eaten by insects, they are amazing chemical factories with the ability to synthesize elaborate compounds from a few abundant ingredients (carbon dioxide, a little nitrogen, and some trace minerals). Among these thousands of compounds are poisons that discourage herbivores, whereas other trees make compounds that disrupt insect behavior. The large, white, crepe-like catalpa flowers are pollinated during the day by bumblebees and at night by large moths. The floral nectar contains a compound, catalposide, that does not affect these legitimate pollinators but will cause behavioral abnormalities, regurgitation, and paralysis in smaller, non-pollinating insects that try to steal the nectar. Catalposide is now being investigated as a potential medicinal compound for humans.

Trunk

Fruits

Seeds

All photos are
of *C. speciosa*

*Acts of
creation are
ordinarily reserved for gods
and poets. To plant a tree,
one need only own a shovel.*
—Aldo Leopold

Key to Commonly Cultivated Catalpas

1. Leaf unpleasant-smelling if crushed, tip shortly pointed, flowers <1.5 in. wide—Southern Catalpa (*C. bignonioides*)
1' Leaf scentless, tip long-pointed, flowers >1.5 in. wide—Western Catalpa (*C. speciosa*)

Ceiba speciosa

SAY-bah spee-see-OH-sah
Ceiba - L., version of South American name
speciosa - L., showy
Malvaceae
Synonym: *Chorisia speciosa*

Brazil and Argentina
Palmate, Alternate
Partially Deciduous, 40–60 ft.

The spectacular floss silk tree is commonly grown throughout Southern California. It can be identified by its showy, pinkish red flowers and swollen green trunk that is often covered with stout prickles. These prickles are likely an adaptation that protects the gray-green, photosynthetic trunk, with its sweet-tasting outer bark, from being girdled by marauding herbivores. The dry fruit is shaped like an avocado and splits open to reveal cottonlike fibers surrounding the seeds. This material, which is closely related to commercial kapok (from *C. pentandra*), has been used in the manufacture of life jackets for the U.S. Navy and by South American Mataco Indians to create arrow-proof vests.

Mature fruit
filled with silky
seed hairs

How can I continue to be astonished by the plane trees lining the avenue—their ubiquity is a disservice to them. We appreciate trees only when they have disappeared, and that is why townspeople dote on them.
—Francis Hallé

Angiosperm: Eudicot: Malvales : **69**

Celtis spp.   Hackberries

SELL-tiss

Celtis - Classic L. name

Cannabaceae

Northern Hemisphere, Tropics

Simple, Alternate

Deciduous, 40–60 ft.

The hackberries are ostensibly similar to elms, but they differ in having gray trunks, three main veins arising from the bases of their leaf blades, and pea-sized, cherry-like fruits. The 60 or so species of *Celtis* are members of the cannabis family, along with marijuana (*Cannabis sativa*) and hops (*Humulus lupulus*). These deep-rooted, tough, and unassuming trees are regularly grown on streets and in parks of cooler interior California cities. In late fall, when their leaves have turned a soft yellow, the ripe fruits are filled with a thin yet tasty flesh.

Smooth, gray, warty
bark of *C. sinensis*

C. australis
leaves & fruits

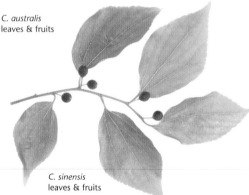

C. sinensis
leaves & fruits

C. australis mature fruits

Key to Commonly Cultivated Hackberries

1. Leaf undersides uniformly fuzzy and felty over the whole surface, leaf margins conspicuously toothed to well below middle, each margin with >20 teeth—European Hackberry (*C. australis*)

1' Leaf undersides hairless or with hairs only along veins, leaf margins smooth (entire) or with teeth only towards the tip, each margin with <16 teeth

  2. Leaves egg-shaped, widest in the middle; bark smooth, without corky outgrowths—Chinese Hackberry (*C. sinensis*)

  2' Leaves narrowly elongate, widest in the basal half, often tapered to an acute tip; bark with prominent corky warts and ridges

    3. Leaves paler beneath; fruit >½ in. diameter, purple when mature—Common Hackberry (*C. occidentalis*)

    3' Leaves same color on both sides, fruit <½ in. diameter, brownish orange or red when mature—Sugarberry (*C. laevigata*)

It is well that you should celebrate your Arbor Day thoughtfully, for within your lifetime the nation's need of trees will become serious. We of an older generation can get along with what we have, though with growing hardship; but in your full manhood and womanhood you will want what nature once so bountifully supplied and man so thoughtlessly destroyed; and because of that want you will reproach us, not for what we have used, but for what we have wasted.
—*Theodore Roosevelt,*
  1907 Arbor Day Message

Ceratonia siliqua

Carob Tree

sair-uh-TOE-nee-ah sill-EE-kwah
Keratonia - Gr., carob tree
siliqua - with curved pods
Fabaceae

Northeastern Africa
Pinnate, Alternate
Evergreen, 40 ft.

T he beautiful, highly drought-tolerant carob tree can be recognized by its shiny, leathery, compound leaves with pairs of nearly round leaflets and by its shiny, lavender-brown, flattened legumes, found only on female trees. This species has had a long history of use by humans in the Middle East and the Eastern Mediterranean. The fruits, which have a pulpy texture on the inside, are high in sugar and protein. Pick up a fallen fruit and bite into the inner flesh to taste carob. The seeds, which are exceptionally uniform in size and weight (about 5 seeds per gram), were the original jeweler's carat weight. Today, the fruit is processed into flour that is used as a chocolate substitute in candy and baked goods, as well as a stabilizer and texturing agent in many cosmetics, pharmaceuticals, and processed foods.

Throughout the ages, humankind has looked to the trees to feed not only the flesh, but the spirit. —George Nakashima

Fruit

Male and female inflorescences

Cercis canadensis   Eastern Redbud

SIR-siss ka-nah-DEN-siss
Kerkis - Gr., weaver's shuttle
canadensis - L., from Canada
Fabaceae

Eastern North America
Simple, Alternate
Deciduous, 20–35 ft.

T he eastern redbud is the largest, fastest growing, and most tree-like of the ten species of *Cercis*. The other commonly grown species is the California native, western redbud (*C. occidentalis*). Redbud leaves have two halves that move independently in response to changes in light. These halves are attached to the tip of the leaf stalk at a swelling called a pulvinus (see *Albizia* discussion). Some authorities believe this highly modified leaf is actually a remnant single, terminal leaflet of an evolutionarily ancestral pinnately compound leaf, a common characteristic of the legume family. In early spring, the mauve-pink redbud flowers emerge directly from the woody parts of the trunk and larger branches, a phenomenon, called cauliflory, found mostly in tropical trees. These distinctive characteristics make redbuds unique among temperate trees. Another redbud cultivated in California is the Judas tree (*C. siliquastrum*, native to the Mediterranean), from which the biblical figure Judas Iscariot supposedly hung himself after betraying Christ.

Eastern Redbud leaf
with short pointed tip

Pulvinus

Western Redbud leaf
with rounded and
notched tip

People who will not sustain trees will soon live in a world which cannot sustain people. —Bryce Nelson

Fruits

Seeds

The maroon-leaved
'Forest Pansy' cultivar

Flowers showing
cauliflory

Chionanthus retusus   Chinese Fringe Tree

kye-owe-NAN-thuss ray-TOO-suss
Chion - Gr., snow; *anthos* - Gr., flower
retus - L., with a rounded tip (leaves)
Oleaceae

China, Korea, Japan, Taiwan
Simple, Opposite
Deciduous, 20–30 ft.

C hinese fringe tree is a relatively recent introduction to California that has become a widely popular urban tree. Its current popularity is understandable after seeing one in full bloom. They develop large clusters of four-petaled, elegant spring flowers, especially in inland areas with winter chill and summer heat. The related fringe tree, *C. virginicus*, from the southeastern U.S., is also occasionally grown in California.

We find our most soothing companionship in trees. We lean against them and they never betray our trust.
—*Oliver Wendell Holmes*

×*Chitalpa tashkentensis* Chitalpa

chi-TALL-pah tosh-kuh-TEN-siss

Chitalpa - L., combination of *Chilopsis*
   and *Catalpa*

tashkentensis - for Tashkent, Uzbekistan

Bignoniaceae

Hybrid
Simple, Opposite, or Whorled
Deciduous, 30 ft.

he chitalpa is a world traveler. It's a horticulturally created hybrid between the southern catalpa (*Catalpa bignonioides*), from western Florida to Louisiana, and the desert willow (*Chilopsis linearis*), native to arid parts of the southwestern United States. These trees were first crossed in Tashkent, the capital city of Uzbekistan, in the 1960s, and eventually became widely grown in the U.S. Chitalpa grows rapidly to about thirty feet, producing profuse white to pink flowers in summer and fall. Like many unlikely hybrids between distantly related parents, chitalpa is sterile, and no fruits are formed after the flowers drop.

Planting a tree gives birth to life and hope. Planting the right tree is even better. —Jeff Reimer

Cinnamomum camphora

Camphor Tree

sin-uh-MOE-mum kam-FOR-ah
Kinnamon - Gr., cinnamon
camphora - L., camphor
Lauraceae

China, Taiwan, Japan
Simple, Alternate
Evergreen, 75 ft.

Fruits

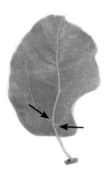

Leaf with three main
veins and two glands

Many California streets are lined with long-lived and stately camphor trees that often create a deeply shaded, overarching tunnel. These broad-leaved evergreens can be recognized by their fissured bark, stout branches, and pale green, glossy leaves that smell of camphor when crushed (camphor oil comes from distilled wood chips). In Japan, where camphors are considered sacred, they have wider trunks than any other tree. So it comes as no surprise that the camphors' aggressive surface roots and ever-widening trunks tend to lift sidewalks, heave curbs, and consume walkways. As their assault on urban infrastructure has become more obvious in recent years, they have become less favored as street trees; fortunately, camphors are still widely planted as lawn trees and on wide parkways throughout California. The bark from the closely related tropical tree *Cinnamomum zeylanicum* is the source of cinnamon sticks and ground cinnamon.

In cities we are never slow to squabble, but when it comes to trees we can agree on two things: we love having them around, and it's a shame we don't know them better. —Arthur Plotnik

Cornus spp.

KOR-nuss
Cornus - L., for Cornelian cherry
Cornaceae

Dogwoods

North Temperates
Simple, Opposite
Deciduous and Evergreen, 20–30 ft.

There are 60 or so species of dogwoods, and these trees are the name-sakes of subdivisions, shopping malls, and festivals throughout the eastern U.S. The most planted in California, frequently in cooler, inland climates, are flowering dogwood (*Cornus florida*), evergreen dogwood (*C. capitata*), Kousa dogwood (*C. kousa*), Cornelian cherry (*C. mas*), the native western dogwood (*C. nuttallii*), and hybrids between these species. You can employ a dogwood to perform two mediocre tricks. Trick number one involves the major leaf veins, which are arranged like the longitudinal lines on a globe and are lined with a latex-like substance that forms fine yet strong strands when pulled apart. Tear a leaf in two, but be careful to leave the two halves connected by the nearly invisible vein strands. Now one half will magically be suspended in mid air, seemingly connected only by mystical ether to the other half held in your hand. If you want to impress your audience, don't attempt this trick on anyone older than four. Trick number two takes advantage of the fact that true dogwood flowers are actually small, yellowish green structures on the inside of the large, petal-like, modified leaves (bracts). At an afternoon party, bet the host on the true color of the flowers on his or her dogwood tree. This trick works best if the host is slightly inebriated. Deploying both tricks at the same party is not advised.

C. florida fruits

C. capitata fruit

C. florida
fall leaf color

The tree which moves some to tears of joy is in the eyes
of others only a green thing that stands in the way.
—William Blake

C. florida flowers

76 : *Angiosperm: Eudicot: Cornales*

Corymbia citriodora   Lemon Scented Gum

koe-RIM-bee-ah sih-tree-oh-DOR-ah
Corymbia - L., a corymb-like inflorescence
citriodora - L., citrus-scented
Myrtaceae
Synonym: *Eucalyptus citriodora*

Northeastern Australia
Simple, Alternate
Evergreen, 75–100 ft.

Juvenile leaves

A beautiful Australian import, the lemon scented gum has an entirely smooth trunk that varies in color from powdery white to coppery pink and pale yellow. While older layers of thin bark are being shed, the trunk can become multicolored and mottled. Save the occasional dimple, there is often no evidence of bygone, self-pruned, lower branches up to nearly half the tree's height. The gracefully contorted trunk, open crown of drooping foliage, and skin-like wrinkles below branch insertions are reminiscent of trees envisioned by Dr. Seuss. When crushed, the bright green, narrow leaves smell of citronella and lemons, and an oil distilled from them is used in perfume manufacturing and as a mosquito repellent. Although the leaves are often too lofty to collect and smell, seeds fallen from high in the canopy germinate under trees in irrigated urban plantings. Search and you will be rewarded by saplings with bristly, glandular, fragrant juvenile leaves.

Fruits ——————

Flower buds ———

Trees are poems that Earth writes upon the sky. We fell them down and turn them into paper, that we may record our emptiness.
—Kahlil Gibran

Corymbia ficifolia   Red Flowering Gum

koe-RIM-bee-ah fiss-sih-FOH-lee-ah

Corymbia - L., a corymb-like inflorescence

ficifolia - L., Ficus-like leaves

Myrtaceae

Synonym: *Eucalyptus ficifolia*

Southwest Australia

Simple, Alternate

Evergreen, 20–50 ft.

Mature fruits

All scientific names are based on hypotheses and can change as new information about the evolutionary relationships between species comes to light. Before 1995, the 100 or so species of *Corymbia* were treated as members of the closely related genus *Eucalyptus*. Taxonomic studies using molecular biology helped scientists realize that the red flowering gum and its close relatives, including the lemon scented gum, are more closely related to *Angophora* trees than they are to *Eucalyptus*. This finding warranted placement in their own genus. Regardless of the name we give it, this species produces an impressive summer display of crimson, vermilion, pink, or orange flowers, often covering the entire crown. The variation in flower color may be influenced by hybridization with the closely related, white-flowered Marri (*C. calophylla*). The red flowering gum can also be identified by its stout trunk, fibrous bark, and dark green, shiny leaves that are lighter on the underside. Although the red flowering gum is very rare in the wild, growing only in scattered stands in the far southwestern corner of Australia, it is one of the most common ornamental eucalypts (a common name that refers to all *Eucalyptus* relatives) in the world.

Variation in flower color

Evolutionary relationships

— *Corymbia*

— *Angophora*

— *Eucalyptus*

Sure, you can name a tree, categorize it, safely identify it. But that tree exists, living the fullness of its quiet life, even if in its long history no man ever stood before it and labeled it. It knows itself already and mysteriously encounters the sun each day, nameless. —Ivan M. Granger

kruh-TEE-gus
Kratos - Gr., strength, referring to the wood
Rosaceae

North Temperates
Simple, Alternate
Deciduous, 25 ft.

Washington Hawthorn
(*C. phaenopyrum*)

Carriere Hawthorn (*C. × lavallei*)

Crataegus is a large and bewildering genus of about 140 species of thorny, deciduous small trees and shrubs mostly from central and eastern North America, Europe, and Asia. The three most commonly grown species in California are the English hawthorn (*C. laevigata*), the Washington hawthorn (*C. phaenopyrum*), and the Carriere hawthorn (*C. × lavallei*). They are prized for their profuse, white to red, spring or early summer flowers, autumn leaf color, and bright red, persistent, pea-sized fruits that are thoroughly relished by birds. These fruits (called "haws") are cooked into jellies and jams, the jagged leaves are steeped into a medicinal tea for treating hypertension and heart disease, and the exceptionally hard wood is crafted into tool handles and walking sticks. Hawthorns make dense growth and hostile thorns that were formerly fashioned into fishing hooks. They can create a menacing and impenetrable hedge and were at one time planted ubiquitously to delineate pastures and property lines throughout Northern Europe. "Haw" is derived from an Old English word for "hedge."

English Hawthorn
(*C. laevigata*)

C. phaenopyrum
leaf and thorn

A scientific interest in at least certain features of our natural environment, as for example the trees, directs one to useful and agreeable intellectual activity. —Willis Linn Jepson

Washington Hawthorn (*C. phaenopyrum*)

Cupaniopsis anacardioides   Carrotwood Tree

koo-pan-ee-OP-sis an-ah-car-dee-OY-deez
Cupania - Francesco Cupani (1657–1710);
    opsis - Gr., likeness
anacardioides - L., like *Anacardium*
Sapindaceae

Northeastern Australia
Pinnate, Alternate
Evergreen, 40 ft.

There is no perfect tree. Each has its virtues and faults, and carrotwood is no exception. It is praised for the deep shade provided by its tidy crown of shiny, dark green leaves and for its tolerance of difficult growing conditions, and it is criticized for its weediness and the mess caused by abundant fruit dropping from mature trees. The carrotwood fruit forms in spring and early summer. Initially, it looks like a large garbanzo bean, but later it develops into an orange, three-parted capsule, splitting open to expose three shiny black seeds, each with a fleshy, orange-red covering. These seeds are attractive to and dispersed over great distances by birds, and because this tree grows well in heat and even salty soil, it has become a noxious weed in southern Florida's marshlands and mangrove and cypress swamps. Carrotwood trees dispersing and reproducing on their own have recently been observed in coastal areas of Southern California—a possible harbinger of future weediness there.

Flowers

Fruit and seeds

Leaf

What is a weed?
A plant whose
virtues have not
been discovered.
—Ralph Waldo
Emerson

Eriobotrya spp.

air-ee-oh-BAH-tree-ah

Erion - Gr., wool; *botrys* - Gr., a cluster of grapes

Rosaceae

China, Taiwan, Japan

Simple, Alternate

Evergreen, 10–30 ft.

E. japonica fruits

E. deflexa fruits

Bronze loquat (*Eriobotrya deflexa*), a densely branched, evergreen tree with lush foliage, is bronze and red when young, eventually turning shiny and dark green with age. The small, purplish brown fruits are barely edible, but the closely related edible loquat (*E. japonica*) is grown almost as commonly in warmer parts of California. Edible loquat's larger, yellow, juicy, acidic fruits can be eaten raw and made into jellies and jams. The fruits of loquats are pomes, similar to apples and pears; the fleshy part of the fruit is derived from the basal portion of the sepals and petals. Look closely at a loquat fruit and you will notice remnants of the five sepals on the end opposite the stem.

Trees are sanctuaries. Whoever knows how to speak to them, whoever knows how to listen to them, can learn the truth. They do not preach learning and precepts, they preach, undeterred by particulars, the ancient law of life.
—Herman Hesse

E. deflexa

E. deflexa flowers

Erythrina spp.

air-ith-RYE-nah

Erythros - Gr., red

Fabaceae

Tropical Americas, Asia, Africa

Pinnate, Alternate

Deciduous, 10–40 ft.

Most of the hundred or so species in the tropical genus *Erythrina* have scarlet red to deep orange, scentless, nectar-filled flowers. Many plants in which pollination by birds has evolved have scentless, reddish flowers—most birds have no sense of smell, and red is invisible to many insects yet conspicuous to birds. Coral trees produce striking red and black seeds. Their leaves are pinnately compound with only three leaflets (trifoliate).

E. coralloides flowers and bark

E. × *sykesii*

E. speciosa prickles on stem

E. × *sykesii* flower development

Banner petal

Banner petal

E. crista-galli flowers

No town fails to be beautiful, though its walks are gutters and its houses hovels, if venerable trees make magnificent colonnades along its streets. —Henry Ward Beecher

E. caffra

E. caffra flowers

Tropical Americas, Asia, Africa
Pinnate (three leaflets), Alternate
Deciduous, 20–60 ft.

Many types of coral trees are grown in Southern California, but two species are more widespread than the rest. The South African coral tree *(E. caffra)*, from eastern South Africa, loses its leaves in winter and then blooms shortly thereafter. The most cold-tolerant of the coral trees, cockspur coral tree *(E. crista-galli)*, from South America, blooms in late spring or summer.

E. caffra smooth bark

E. caffra fruit and seeds

E. crista-galli rough bark

Key to California's Commonly Cultivated *Erythrina*

1. Mature leaves covered with fine hairs, especially on the lower surface and when young
    2. Bark corky; upper petal (banner) short and broad, reflexed, exposing the stamens, standing up from other flower parts—Broad Leaved Coral Tree (*E. latissima*)
    2′ Bark smooth; upper petal (banner) long and narrow, tube-like, only slightly reflexed, enclosing the stamens, folded around other flower parts—Pink Coral Tree (*E. speciosa*)
1′ Mature leaves hairless
    3. Leaflets ovoid (egg-shaped) or rhomboid (diamond-shaped), widest in the middle; terminal leaflet often longer than wide
        4. Leaf margins flat; prickles present on some or all leaves; seeds brown; inflorescence born on shoot tips—Cockspur Coral Tree (*E. crista-galli*)
        4′ Leaf margins wavy; prickles absent or minute on leaves; seeds red, orange, or buff; inflorescence borne in leaf axils—Brazilian Coral Tree (*E. falcata*)
    3′ Leaflets triangle-shaped, widest toward the base; terminal leaflet often similar in length and width
        5. Leaves without prickles
            6. Banner petal short and broad, reflexed, exposing the stamens, standing up from other flower parts—South African Coral Tree (*E. caffra*)
            6′ Banner petal long and narrow, tube-like, only slightly reflexed, enclosing the stamens, folded around other flower parts
                7. Fruits never formed; keel petals free, ~¼ the length of the banner petal; calyx two-lipped—Sykes Coral Tree (*E. × sykesii*)
                7′ Fruits often present on mature trees; keel petals fused, ~⅛ the length of the banner petal; calyx cup-shaped—Transvaal Coral Tree (*E. lysistemon*)
        5′ Leaves bearing prickles
            8. Bark rusty, orange, or cinnamon; banner petal more than 6 times longer than wide—Naked Coral Tree (*E. coralloides*)
            8′ Bark gray, green, or brown; banner petal less than 5 times longer than wide
                9. Calyx 5-toothed; inflorescences borne above foliage and canopy; mature flowers downturned on inflorescence, orangish red—Natal or Dwarf Coral Tree (*E. humeana*)
                9′ Calyx two-lipped, without teeth; inflorescences borne within the crown, not elevated above canopy; mature flowers erect on inflorescence or only slightly downturned, crimson-red—Bidwill's Coral Tree (*E. × bidwillii*)

Eucalyptus spp.

yoo-kuh-LIPP-tuss

Eu - Gr., well; *kalypto* - Gr., covered

Myrtaceae

*What peace comes to those aware of the voice
and bearing of trees! —Cedric Wright Henderson*

Coral Gum (*E. torquata*) flowers

Eucalypts are the most widespread of all California's cultivated trees. Nary an urban or suburban skyline exists in western California without the canopy of a eucalypt somewhere in the distance. They were introduced to California from Australia in the 1850s to be grown as horticultural oddities for the nursery trade, then later as promising forestry trees and possible saviors during a forecasted timber drought. By the early 1900s, blue gum (*Eucalyptus globulus*) was being extensively planted for lumber, pilings and posts, fuel wood, medicinal products, tannin, oil, windbreaks, and as a street and park tree. However, as the California forestry and fuel economy evolved, and the inferior quality of young blue gum wood was discovered, most of the plantations remained uncut, and parts of the state are now burdened with the ecological legacy of this vast unharvested crop. In the regions where they are now conspicuous landscape features, they are either admired as aesthetically valuable heritage trees and monarch butterfly habitat, or demonized as America's largest weeds.

Narrow-Leaf Peppermint (*E. nicholii*)

Red Iron Bark (*E. sideroxylon* 'Rosea')
blue leaves and pink flowers

Australia
Simple, Opposite and Alternate
Evergreen, 10–200 ft.

The 700 or so species of eucalypts grow natively almost exclusively in Australia. They exhibit a great range of adaptation to different moisture conditions and rival other large tree genera, such as figs (*Ficus* spp.), pines (*Pinus* spp.), and oaks (*Quercus* spp.), in the diversity of their mature tree sizes. Species range from small, multi-stemmed shrubs (often called "mallees") to some of the tallest and largest forest trees on Earth. In fact, the tallest flowering plant reported in North and South America is a blue gum towering over 240 feet above the beach off the coast of California, on Santa Cruz Island.

More than 300 different species of eucalypts have been grown in California, mainly as experimental forestry, park, garden, and specimen trees. They are grown frequently

Silver Dollar Gum (*E. polyanthemos*) canopy

along roads and highways and have the appealing characteristics of drought tolerance, beautiful evergreen foliage, striking bark, and in some species, showy flowers.

Sugar Gum (*E. cladocalyx*)

Blue Gum (*E. globulus*) grove

*The wonder is that we
can see these trees and
not wonder more.*
—*Ralph Waldo Emerson*

Eucalyptus spp.

Eucalypt bark, leaves, and reproductive structures are greatly varied, and you may need to examine each in order to identify a tree with certainty. Many species retain dead bark year after year, giving rise to a trunk covered in a hard, weathered, outer layer, such as the red iron bark tree (*E. sideroxylon*). Others lose old layers of bark annually, resulting in a completely smooth trunk, such as that of the white iron bark (*E. leucoxylon*). Eucalypts have two leaf forms: juvenile leaves are commonly covered in whitish wax, attached directly to the stem in opposite pairs, and oriented horizontally, whereas adult leaves tend to be stalked, attached singly, hanging vertically, and shaped like spearheads. Some species, such as the silver mountain gum (*E. pulverulenta*), which is commonly used in cut flower arrangements, indefinitely retain their juvenile leaves.

Bud cap

Coral Gum (*E. torquata*) flowers and fruits developing in series

White Iron Bark (*E. leucoxylon*)

Narrow-Leaf Peppermint (*E. nicholii*)

Red Gum (*E. camaldulensis*)

Red Iron Bark (*E. sideroxylon*)

Swamp Mahogany (*E. robusta*)

Red Flowering Gum (*Corymbia ficifolia*)

As in most genera, the defining aspects of eucalypts are in their reproductive structures. The flowers of only a small number of species, such as the blue gum, develop individually in leaf axils; in most species they develop in umbrella-shaped clusters. Individual clusters, usually with three, seven, or eleven or more flowers, may develop singly in leaf axils, as in red gum (*E. camaldulensis*), or in highly branched structures at shoot tips, as in the silver dollar gum (*E. polyanthemos*).

The name "Eucalyptus" is derived from the Greek words *eu* ("well") and *kalyptos* ("covered"), referring to a leathery, dome-shaped bud cap derived from fused sepals and petals. This bud cap falls off as the flowers open to reveal their most conspicuous component: many white, or sometimes colorful, stamens. Eucalypt flowers, which are mostly insect-pollinated, develop into woody, cup-shaped fruits that open at the top, shedding tiny, wind-dispersed seeds.

Red Cap Gum (*E. erythrocorys*) with yellow stamens

Karri (*E. diversicolor*) with bud cap falling off

Blue Gum (*E. globulus*) with bud cap falling off

Red Gum
(*E. camaldulensis*)
fruits

White Iron Bark
(*E. leucoxylon*) fruits

Silver Dollar Gum
(*E. polyanthemos*) fruits

*Keep a green tree
in your heart and
perhaps a singing
bird will come.*
—*Chinese Proverb*

Swamp Mahogany (*E. robusta*) leaves, lighter on the underside

Argyle Apple (*E. cinerea*) juvenile leaves and fruits

He that plants a tree loves others besides himself.
—*Thomas Fuller*

Key to California's Commonly Cultivated Eucalypts

1. Leaves lighter green on the underside
  2. Bark rough, hard, thin, flaky; fruit ¾ in. (2 cm) diameter or more—Red Flowering Gum (*Corymbia ficifolia*)
  2' Bark rough, soft, thick, fibrous; fruit ½ in. (1.3 cm) diameter or less—Swamp Mahogany (*E. robusta*)
  2" Bark smooth
    3. Bark with orange blotches; leaves often curved; mature fruit ridged; fruit valves sunken inside fruit—Sugar Gum (*E. cladocalyx*)
    3' Bark bluish gray; leaves straight; mature fruit smooth; valves of fruit exserted
      4. Valves 4 or 5, curved inward; buds and fruit often glaucous—Rose Gum (*E. grandis*)
      4' Valves usually 4, erect; leaves straight; buds and fruit not glaucous—Sydney Blue Gum (*E. saligna*)
1' Leaves the same color on both sides; *and* bark rough, furrowed, retained on trunk and limbs (*Note*: there is a third option 1")
  5. Leaves silver or bluish silver in color
    6. Leaves attached oppositely; flower buds and fruit grouped in threes—Argyle Apple (*E. cinerea*)
    6' Leaves attached alternately, mostly less than ½ in. (1.3 cm) wide at widest point—Narrow-Leaf Peppermint (*E. nicholii*)
    6" Leaves attached alternately, mostly ¾ in. (2 cm) wide or wider
      7. Bark dark brown to black, very rough and deeply furrowed—Red Iron Bark (*E. sideroxylon*)
      7' Bark soft, thick, fibrous; fruit ½ in. diameter or less—Swamp Mahogany (*E. robusta*)
      7" Bark light brown or gray, scraggly; leaves 3 to 4 times longer than broad; flowers pink—Coral Gum (*E. torquata*)
  5' Leaves green
    8. Leaves mostly less than ½ in. (1.3 cm) wide at widest point—Narrow Leaf Peppermint (*E. nicholii*)
    8' Leaves mostly ¾ in. (2 cm) wide or wider; bark gray, finely furrowed; all stamens with anthers—Flooded Gum (*E. rudis*)
    8" Leaves mostly ¾ in. (2 cm) wide or wider; bark dark brown to black, very rough and deeply furrowed; some stamens without anthers—Red Iron Bark (*E. sideroxylon*)
1" Leaves the same color on both sides; *and* bark smooth, shedding (sometimes with imperfectly shed rough bark on the basal area of trunk)
  9. Buds and fruit produced singly; leaves over 8 in. (20 cm) long—Blue Gum (*E. globulus*)
  9' Buds and fruit clustered in threes or more; leaves less than 6 in. (15 cm) long, smelling like lemon when crushed—Lemon Scented Gum (*Corymbia citriodora*)
  9" Buds and fruit clustered in threes or more; leaves less than 6 in. (15 cm) long, smelling medicinal or spicy when crushed (but not like lemon)
    10. Leaves silver, bluish silver, or bluish gray in color (covered with wax)
      11. Shrub with several trunks; all leaves opposite, without a leaf stalk—Silver Mountain Gum (*E. pulverulenta*)
      11' Tree with one trunk; some leaves alternate, stalked; buds and fruits in simple clusters of three in leaf axils—Cider Gum (*E. gunnii*)
      11" Tree with one trunk; some leaves alternate, stalked; buds and fruits in branched clusters of seven at shoot tips—Silver Dollar Gum (*E. polyanthemos*)
    10' Leaves green, olive green, or dark green (not waxy)
      12. Leaves ovoid (egg-shaped) or round, barely twice as long as wide; buds and fruit fused into a spherical cluster the size of a small fist—Spider Gum (*E. conferruminata*)
      12' Leaves shaped like a long spearhead, 3 to 4 times longer than wide; buds and fruit not fused
        13. Buds and fruit in clusters of five or more (usually seven)
          14. Bud hemispheric, with a distinct beak; branches often drooping—Red Gum (*E. camaldulensis*)
          14' Bud horn-shaped or conical, not beaked; branches often steeply ascending—Forest Red Gum (*E. tereticornis*)
        13' Buds and fruit always in clusters of three
          15. Bark shedding in long slender ribbons; buds with a short stalk; all stamens with anthers—Ribbon Gum (*E. viminalis*)
          15' Bark shedding in plates or patches; buds borne on a long slender stalk; some stamens without anthers—White Iron Bark (*E. leucoxylon*)

What did the tree learn from the earth to be able to talk with the sky?
—Pablo Neruda

Buds and Fruits for California's Commonly Cultivated Eucalypts

Red Gum
(*Eucalyptus
camaldulensis*)

Blue Gum
(*Eucalyptus globulus*)

Manna Gum
(*Eucalyptus viminalis*)

Sugar Gum
(*Eucalyptus cladocalyx*)

Spider Gum
(*Eucalyptus
conferruminata*)

Coral Gum
(*Eucalyptus
torquata*)

White Iron Bark
(*Eucalyptus
leucoxylon*)

Silver Dollar
Gum (*Eucalyptus
polyanthemos*)

Sydney
Blue Gum
(*Eucalyptus
saligna*)

Swamp Mahogany
(*Eucalyptus robusta*)

Lemon Scented Gum
(*Corymbia citriodora*)

Narrow-Leaf
Peppermint
(*Eucalyptus
nicholii*)

Argyle Apple
(*Eucalyptus cinerea*)

Red Flowering Gum
(*Corymbia ficifolia*)

Silver Mountain
Gum (*Eucalyptus
pulverulenta*)

Cider Gum
(*Eucalyptus
gunnii*)

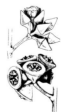

Flooded Gum
(*Eucalyptus rudis*)

Red Iron Bark
(*Eucalyptus
sideroxylon*)

Ficus spp.

FYE-kuss

Ficus - L., name for the edible fig

Moraceae

All fig species, including the only one regularly eaten by humans (*Ficus carica*), have common characteristics. When damaged they bleed milky sap, and the tip of each stem is covered by a pair of leaflike sheaths (called stipules) that fall as new leaves unfurl. Their minute flowers are borne on the inside walls of a hollow stem, a structure called a syconium, that eventually develops into the ripe fig. Also, each of the 800 fig species has a corresponding wasp species that pollinates it. Female wasps squeeze inside the hole at the tip of the fig and pollinate the tiny flowers while using the chamber as a nursery for their young. If this makes you never want to eat another fig, relax—most commercial figs develop without wasp pollination.

There are nearly 800 species of figs (*Ficus*). Found mostly in tropical regions, they range from small climbing vines to some of the largest tropical trees in the world—including one individual banyan (a common name for figs that form extensive aerial roots that become additional trunks) in Calcutta, India, that covers nearly four acres and has over fifteen hundred subsidiary trunks formed from aerial roots.

F. *auriculata* fruits

F. *rubiginosa*

F. *macrophylla*

Those who fancy that humans are superior to the rest of nature often use "tree-hugger" as a term of ridicule, as if to feel the allure of trees were a perverted form of sensuality or a throwback to our simian ancestry.
—*Scott Russell Sanders*

Tropics and Subtropics
Simple, Alternate
Evergreen, 20–200 ft.

F. elastica

F. carica

F. rubiginosa

F. microcarpa

F. macrophylla

F. benjamina

Edible Fig (F. carica) shoot tip showing
deciduous stipules (note arrows)

Rusty leaf fig (F. rubiginosa)
branches and syconia

Large fig of
F. auriculata

Key to California's Commonly Cultivated *Ficus*

1. Plant a climbing vine—Creeping Fig *(F. pumila)*
1' Plant a tree or large shrub
  2. Leaves 3- to 5-lobed, deciduous, with 3 main veins; tree cultivated agriculturally—
     Edible Fig *(F. carica)*
  2' Leaves not lobed, evergreen, with 1 main vein; tree cultivated ornamentally
    3. Terminal bud covered by 1 (often reddish) sheath (stipule)—Rubber Tree
       (F. elastica)
    3' Terminal bud covered by 2 sheaths (stipules)
      4. Leaves covered with very fine rust-colored hairs (at least when young)
        5. Mature trees with large buttress roots; leaves 2.5 to 5 in. wide, figs ¾ to
           1 in. wide—Moreton Bay Fig *(F. macrophylla)*
        5' Mature trees without large buttress roots, leaves 1.5 to 2.5 in. wide,
           figs ¼ to ½ in. wide—Rusty Leaf Fig *(F. rubiginosa)*
      4' Leaves hairless
        6. Leaf veins inconspicuous, very numerous, leaf with a long, pointed tip—
           Weeping Fig *(F. benjamina)*
        6' Leaf veins conspicuous, usually <10, leaf without a long tip—Indian Laurel
           Fig *(F. microcarpa)*

Ficus macrophylla

Moreton Bay Fig

FYE-kuss mak-row-FILL-ah

Ficus - L., name for the edible fig

macrophylla - Gr., large leaves

Moraceae

Eastern Australia
Simple, Alternate
Evergreen, 50–75 ft.

T he largest and most remarkable figs in California are the Moreton Bay figs. They have massive buttressing roots, sweeping canopies, and woody, aerial roots that flow and twist downward to the soil. Many spectacular specimens planted in the 1800s persist in Southern California.

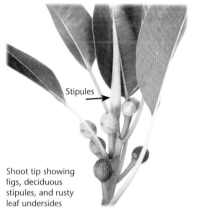

Stipules

Shoot tip showing figs, deciduous stipules, and rusty leaf undersides

Old trees in their living state are the only things that money cannot command.
—Walter Savage Landor

California's Notorious Moreton Bay Figs

Santa Barbara Train Station
34.413586, -119.694006
Planted 1876

Balboa Park, San Diego
32.73275, -117.147467
Planted 1914

Mission Park, Ventura
34.28035, -119.298219
Planted 1874

St. Luke's Hospital, San Francisco
37.747481, -122.420358
Planted 1905

Fairmont Miramar Hotel, Santa Monica
34.0177, -118.501383
Planted 1889

Owens Tree, Glendora
34.126922, -117.866381
Planted 1905

Beverly Hills Fig
34.072256, -118.403869
Planted 1910

La Mesa Drive Trees, Santa Monica
34.0489, -118.494264
Planted 1920

Ficus microcarpa

Indian Laurel Fig

FYE-kuss mye-kroe-CAR-pah
Ficus - L., name for the edible fig
microcarpa - Gr., small-fruited
Moraceae

Indo-Malaysia
Simple, Alternate
Evergreen, 40–60 ft.

The most commonly planted ornamental fig in California, especially as a large street tree in Central and Southern California cities, is the Indian laurel fig. Its giant rounded crown of shiny, dark green leaves is supported by sinuous, smooth, gray branches and a trunk that often originates from an impossibly small square in a city sidewalk. The roots of older trees tend to become aggressive, damaging sidewalks and foundations, a character that has led to an unfortunate decline in the popularity of this species in some California cities.

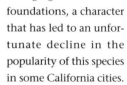

Shoot tip showing paired figs

A tree is beautiful, but what's more, it has a right to life; like water, the sun and the stars, it is essential. Life on earth is inconceivable without trees. —Anton Chekhov

Fraxinus spp.

FRACK-sih-nuss

Fraxinus - L., ash tree

Oleaceae

The 65 or so species of ashes are mostly deciduous and most are from the north temperates. They are most easily recognized by their opposite, pinnately compound leaves, their inconspicuous, wind-pollinated, greenish flowers, and their narrow, single-seeded, winged fruits. Ashes, which are usually fast-growing, are planted as street, shade, and park trees in many cities. Fine-grained, cream-colored ash wood is strong yet elastic and therefore excels in such uses as tool handles, oars, and baseball bats.

Central leaf axis

Leaflet

Node

Typical opposite, pinnately compound leaves

F. angustifolia 'Raywood' fall color

F. uhdei as street trees

We have relatively narrow means by which to approach a tree.

It may be in the way, or it may have ornamental value. For those who deal in lumber it will have another; and most people do not know its name. We very rarely assume that any such silent, faithful, available plant would be able to draw any more out of us than a tacit acceptance.
—John Fay

F. velutina fall color

F. uhdei, evergreen in midwinter

North Temperates
Pinnate, Opposite
Evergreen or Deciduous, 30–80 ft.

The most commonly grown ashes in California are Modesto ash (*Fraxinus velutina* 'Modesto'), shamel ash (*F. uhdei*), and raywood ash (*F. angustifolia* 'Raywood'). California is also home to four native ashes and several other occasionally grown species from the eastern U.S. and Europe. Modesto ash, a cultivar of *F. velutina* that originated in Modesto, California, was fervently planted in many cities in California in the 1940s through the 1960s but has since fallen out of favor due to its susceptibility to a number of diseases and to parasitic mistletoe infestations. Evergreen shamel ash, which is native to southern Mexico, Guatemala, and Honduras, is the most commonly planted ash in Southern California. It can reach a great size and, if planted as a street tree, tends to heave sidewalks. The raywood cultivar of *F. angustifolia* is a more compact tree and is popular as a street tree in much of California. It bears a lacy, rounded crown of slender leaflets that turn a glorious plum purple in the fall.

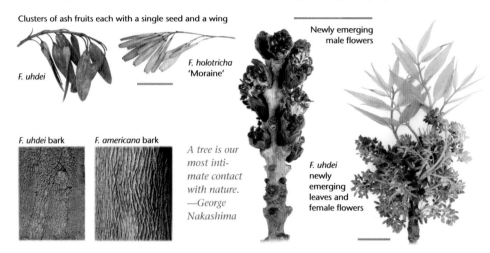

Clusters of ash fruits each with a single seed and a wing

F. uhdei

F. holotricha 'Moraine'

Newly emerging male flowers

F. uhdei bark   *F. americana* bark

A tree is our most intimate contact with nature.
—George Nakashima

F. uhdei newly emerging leaves and female flowers

Key to California's Commonly Cultivated Ashes

1. Leaflets without a stalk (sessile or nearly so)
   2. Leaflets shaped like narrow spearheads, dark green, usually 7 to 9—Raywood Ash (*F. angustifolia* 'Raywood')
   2' Leaflets oblong or egg-shaped, light green, usually 5 to 7—Oregon Ash (*F. latifolia*)
1' Leaflets stalked
   3. Flowers terminal, whitish-green, with petals, emerging after leaves—Flowering Ash (*F. ornus*)
   3' Flowers in leaf axils, greenish, without petals, inconspicuous, emerging with new leaves
      4. Leaf undersides hairy, especially on central leaf axis—Green Ash (*F. pennsylvanica*)
      4' Leaf underside hairless, or with only stiff hairs bordering leaf veins
         5. Leaflets usually 5 (or 3 to 7)—Velvet or Modesto Ash (*F. velutina*)
         5' Leaflets usually 7 (or 5 to 9)
            6. Leaflet margins regularly toothed, underside green, tree evergreen or briefly deciduous, bark scaly—Shamel Ash (*F. uhdei*)
            6' Leaflet margins smooth, wavy, or sparsely toothed, undersides whitish, tree deciduous, bark with interlacing corduroy-like ridges—White Ash (*F. americana*)

gye-JEER-uh par-vih-FLOR-uh
Geijera - J. D. Geijer (1660–1735)
parviflora - L., small-flowered
Rutaceae

Eastern Australia
Simple, Alternate
Evergreen, 30 ft.

Flowers

Typical branching structure

T he Australian willow (known in Australia by the Indigenous name "Wilga") is one of five species of *Geijera* endemic to Australia. It is native to the arid mixed woodlands found in the interior parts of eastern Australia and has all the toughness associated with evolving in a harsh, dry climate. Its deep and noninvasive roots are tolerant of poor soils, compact growing spaces, and long periods of drought. Although the linear, drooping leaves on pendent branches resemble those of a true willow (*Salix* spp.), the Australian willow is actually a member of the citrus family. A clue that partly reveals this relationship is the subtle yet sweet scent of its tiny flowers, like orange trees in bloom. Introduced as a street tree to California by the Saratoga Horticultural Foundation in the late 1950s, this species has a characteristically narrow, upright branching architecture, similar to the Callery pear (see page 136). In its native land, the waxy, aromatic leaves are chewed as a painkiller, smoked for their psychoactive properties during Indigenous Australian ceremonies, and used as sheep fodder.

*I touch trees, as others
might stroke the fenders
of automobiles or finger
silk fabrics or fondle cats.
Trees do not purr, do not
flatter, do not inspire a
craving for ownership or
power. They stand their
ground, immune to merely
human urges. Saplings
yield under the weight of a
hand and then spring back
when the hand lifts away,
but mature trees accept
one's touch without so
much as a shiver.*
—Scott Russell Sanders

Gleditsia triacanthos   Honey Locust

gleh-DIT-see-ah try-ah-KAN-those
Gleditsia - Johann Gottlieb Gleditsch
   (1714–1786)
triacanthos - Gr., three-thorned
Fabaceae

Eastern North America
Pinnate or Bipinnate, Alternate
Deciduous, 40–70 ft.

A native of eastern North America, the durable honey locust is planted on streets in many cities. It stays deciduous longer than any other ornamental tree in California, being mostly leafless by Halloween and rarely with new leaves before Easter. The honey locust can be recognized by its open, lacy crown of bright green, fernlike leaves, often pinnate and bipinnate on the same tree. Its inconspicuous, greenish flower clusters develop into flat, twisted, brown pods containing sweet pulp, hence the common name. Some trees are armed with formidable thorns on the trunk and large branches. However, most cultivars (usually of *G. triacanthos* forma *inermis*) are thornless, and some are pendulous ('Butjoti'), dwarfed ('Nana'), have golden spring foliage ('Sunburst'), or are sterile ('Moraine') and thus produce no messy fruits.

Flowers

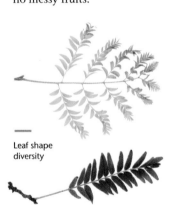

Leaf shape
diversity

'Sunburst' golden spring foliage

Mature fruit

A tree never hits an automobile except in self-defense.
—*American Proverb*

Bark and banched thorns

Grevillea robusta   Silk Oak

gruh-VILL-ee-yah roe-BUS-tah

Grevillea - Charles Francis Greville (1749–1809)

robusta - L., stout

Proteaceae

Eastern Australia

Pinnate, Alternate

Evergreen, 50–60 ft.

ilk oak is not a true oak (*Quercus* spp.), but is a member of the Protea family, a Southern Hemisphere family of showy plants that include the proteas, banksias, and macadamia nut (*Macadamia integrifolia*). Silk oak is the largest of the 360 or so species of *Grevillea*—350 species are endemic to Australia. It is a fast-growing tree with dark green, deeply lobed leaves with silver, silky undersides. Trees growing in warmer parts of California bloom profusely with golden orange, nectar-rich flowers that present their pollen to pollinators not with the extended male stamen typical to most plants, but with a female style. In each of the small flowers, the tip of the style is looped into a pocket full of pollen; as the style uncurls, it offers the pollen to passing insects and other pollinators. Silk oak's pale, pinkish wood, which is reminiscent of oak, is used for furniture and cabinets. This species reproduces vigorously in moist climates and is considered an invasive weed in Hawaii, southern Florida, and other subtropical parts of the world.

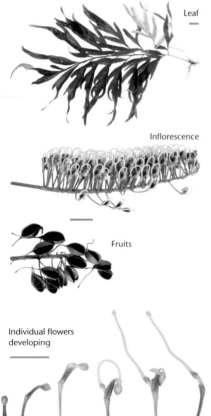

Leaf

Inflorescence

Fruits

Individual flowers
developing

How little we know of our trees, even those casting their friendly shadows across our daily paths. —Maunsell van Rensselaer

Handroanthus spp.

han-droe-ANN-thus

Handro - Oswaldo Handro (1908–1986);

   anthus - flower

Bignoniaceae

Synonym: *Tabebuia* spp.

Central, South America

Palmate, Opposite

Partially Deciduous, 20–50 ft.

rumpet trees are members of the mostly tropical Bignonia family, which is usually characterized by opposite, compound leaves, two-lipped, tubular flowers, and linear capsules filled with winged seeds. This family contains some of the showiest trees in the world, including jacaranda (page 102) and catalpa (page 68). Many trumpet trees are important in their neotropical habitats as producers of durable lumber, imported into the U.S. under the name "ipê." Two species of trumpet trees are regularly grown in warmer parts of California: pink trumpet tree (*H. heptaphyllus*), with light pink to purple flowers, and golden trumpet tree (*H. chrysotrichus*), with maroon-striped, golden flowers. Each spring both species briefly drop their leaves before staging an extravagant show of brilliant flowers. A mature, leafless, trumpet tree in full bloom seems worthy of a call to the local newspaper.

My relations with trees are made simple and calm by my feelings of shedding a universal sadness. Their simple presence is profoundly comforting and calming. —Yildiz Aumeeruddy

H. heptaphyllus
flowers

H. heptaphyllus
leaves and fruit

Leaves

H. chrysotrichus
tree

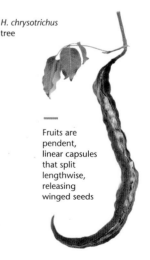

Fruits are
pendent,
linear capsules
that split
lengthwise,
releasing
winged seeds

Heteromeles arbutifolia Toyon

het-er-OH-mih-leez ar-bew-tih-FOE-lee-ah
Heteros - Gr., different; *meles* - Gr., apple
arbutifolia - L., with leaves like *Arbutus*
Rosaceae

California, Oregon, Baja California
Simple, Alternate
Evergreen, 30 ft.

Toyon is a small tree or large shrub native to California, where it grows as a component of chaparral plant communities throughout the western portion of the state. It's the only California native plant widely called by a Native American name (*toyon* is the Spanish adaptation of the southern Bay Area Ohlone word for the plant, *tottcon*). Throughout summer, toyons develop clusters of white flowers, followed in the fall and early winter by pea-sized, bright red fruit. In the past, entire limbs bearing these beautiful fruits were gathered from the wild and sold commercially for decorations at Christmastime (another common name is Christmas berry). This easy-to-grow, adaptable shade-tolerant plant can make an attractive small street tree, focus specimen, or evergreen screen. The winter fruits are an important food source for birds, including cedar waxwings, American robins, and hermit thrushes.

A tree planted without forethought often resembles an afterthought.
—Jeff Reimer

Young flowers

Leaf

Fruit

hye-men-AWE-spor-um FLAY-vum

Hymen - Gr., membrane; *spora* - Gr., seed

flavum - L., yellow

Pittosporaceae

Eastern Australia
Simple, Alternate
Evergreen, 20–40 ft.

Flowers at different stages of maturity

One of California's most sweet-smelling trees, sweetshade has creamy flowers with a delightful fragrance, mostly in the evening, that combines the scents of honey, orange blossoms, frangipani (*Plumeria rubra*), and mock orange (*Pittosporum undulatum*, page 128). These flowers open white, soon turn to deep golden yellow, then burnt orange, and eventually wither to brown. The pollen-producing anthers protrude in the younger white flowers and then wither, only to be followed by a receptive and protruding female stigma in older flowers. This nonsynchronous timing likely evolved in these hermaphroditic flowers because self-pollination was detrimental. Sweetshade has an upright, slender crown of glossy, dark green leaves that are often whorled around branch ends. It should be grown in a protected spot as it can succumb to wind, drought, and other harsh realities of urban tree living. The sweetshade is closely related to *Pittosporum* but differs in having winged, nonsticky seeds.

Mature fruits
with seeds

I think that I shall never see
a billboard lovely as a tree.
Perhaps, unless the billboards fall,
I'll never see a tree at all.
—Ogden Nash

Jacaranda mimosifolia   Jacaranda

jak-uh-RAN-dah mih-moe-sih-FOE-lee-ah
Jacaranda - Brazilian name for the genus
mimosifolia - L., leaves like mimosa
Bignoniaceae

Argentina, Bolivia
Bipinnate, Opposite
Deciduous, 25–45 ft.

A world-renowned flowering tree from the seasonally dry tropics of South America, jacaranda enlivens city streets throughout sub-tropical and warmer temperate parts of the world. This species famously covers Pretoria, South Africa (also known as Jacaranda City), where it has now become illegal to buy or sell a jacaranda, because of the tree's potential to become an invasive weed. California is much drier than eastern South Africa, and jacaranda, one of our most floriferous trees, rarely reproduces on its own here. Jacarandas are scarce in Northern California but common in Southern California, where streets come alive in early summer with clouds of lavender, lilac, halogen blue, and purple. When not flowering, the jacaranda can be distinguished by its fernlike leaves, each with hundreds of tiny leaflets, and by the oval seed pods that split open to release winged seeds. The leaves fall in spring and return in summer as the flowers fade.

Winged seeds and
mature fruit

Because they are primeval, because they outlive us, because they are fixed, trees seem to emanate a sense of permanence. And though rooted in earth, they seem to touch the sky. For these reasons it is natural to feel we might learn wisdom from them, to haunt about them with the idea that if we could only read their silent riddle rightly we should learn some secret vital to our own lives; or even, more specifically, some secret vital to our real, our lasting and spiritual existence. —Kim Taplin

Leaf

Flower

Juglans spp.

Walnuts

JUG-lens
Juglans - L., walnuts
Juglandaceae

North Temperates
Pinnate, Alternate
Deciduous, 20–50 ft.

There are twenty-one species of walnuts worldwide, and two are native to California. The state has a long history of walnut farming, especially in the Sacramento Valley, where there are many English walnut (*J. regia*) orchards. In addition to the English walnut, two other species, Northern California black walnut (*J. hindsii*) and Southern California black walnut (*J. californica*), are occasionally cultivated as street, park, or garden trees.

Key to California's Commonly Cultivated Walnuts
1. Leaflets >1.5 in. wide—English Walnut (*J. regia*)
1′ Leaflets 1 in. or less wide
2. Leaflets generally widest in the middle, tips rounded to a point; leaflet undersides without hairs—Southern California Black Walnut (*J. californica*)
2′ Leaflets generally widest at the base, tips tapering to a point; leaflet undersides with tufts of hairs along the veins—Northern California Black Walnut (*J. hindsii*)

J. hindsii (left) and *J. regia* fruit (right)

We complain and complain, but we have lived and seen the blossom—apple, pear, cherry, plum, almond blossom—in the sun; and the best among us cannot pretend they deserve—or could contrive—anything better.
—J.B. Priestley

J. regia

J. californica fruits

J. hindsii male inflorescences

J. hindsii female flowers

Koelreuteria spp.

kole-roo-TARE-ee-ah

Koelreuteria - Joseph G. Koelreuter (1733–1806)

Sapindaceae

Seeds

The three species in the genus *Koelreuteria* are all elegant trees used often in California's urban landscapes. The most commonly grown is the goldenrain tree (*K. paniculata*), from central China, that can survive to 0°F. It was first introduced into the United States by Thomas Jefferson in 1809 when he received a shipment of seeds from France, where the tree had been introduced earlier from Asia. The trees are revered in their homeland, where they are planted around special gravesites, and the flowers are used medicinally and as a source of yellow dye. The hard, black seeds are also used as beads in religious ceremonies. The two other species of *Koelreuteria,* also found regularly in California, are the Chinese flame tree (*K. bipinnata*), from southwest China, introduced to the U.S. by Santa Barbara's famous plantsman Dr. Francesco Franceschi, and the flamegold tree (*K. elegans*), from Taiwan and Fiji. This is the most frost-tender of the three. All three species have fragrant, bright yellow flowers that emerge in late summer, followed by burgundy to tan-colored, papery fruits that resemble Chinese lanterns.

K. bipinnata
and *K. elegans* leaf   *K. paniculata* leaf

K. bipinnata canopy with
flowers, leaves, and fruits

Key to *Koelreuteria*

1. Leaves pinnately compound
   (once divided into leaflets)
   —*K. paniculata*
1' Leaves bipinnately compound
   (twice divided into leaflets)
   2. Leaflet bases conspi-
      cuously asymmetric,
      petals 5—*K. elegans*
   2' Leaflet bases symmetric
      (or only slightly asym-
      metric), petals 4 (rarely
      5)—*K. bipinnata*

China, Taiwan, Fiji
Pinnate or Bipinnate, Alternate
Deciduous and Evergreen, 20–40 ft.

K. bipinnata flowers

K. paniculata fruits

*You can gauge a
country's wealth,
its real wealth,
by its tree cover.
—Richard
    St. Barbe Baker*

K. bipinnata

The Ten Trees Most Likely to Trip You on the Sidewalk

Sheoak *(Casuarina* spp.*)*—Round, conelike fruits
Floss Silk Tree *(Ceiba speciosa)*—Slippery fallen flowers
Camphor Tree *(Cinnamomum camphora)*—Large roots push up sidewalks
Blue Gum *(Eucalyptus globulus)*—Round fruits
Jacaranda *(Jacaranda mimosifolia)*—Slippery fallen flowers and fruits
Golden Rain Tree *(Koelreuteria* spp.*)*—Seeds like ball bearings
Sweetgum *(Liquidambar styraciflua)*—Round fruits, roots
Southern Magnolia *(Magnolia grandiflora)*—Roots and slippery, waxy leaves
Oaks *(Quercus* spp.*)*—Acorns
Queen Palm *(Syagrus romanzoffiana)*—Round fruits

Lagerstroemia spp.   Crape Myrtle

lah-gur-STROE-mee-ah

Lagerstroemia - Magnus von Lagerström
  (1691–1759)

Lythraceae

China, Japan

Simple, Opposite or Alternate

Deciduous, 10–30 ft.

An array of desirable traits, such as compact size, showy, long-lasting flowers, and striking fall foliage have made the crape myrtle a popular small ornamental tree in California and elsewhere. Many different cultivated varieties and hybrids between the Chinese *Lagerstroemia indica* and the Japanese *L. fauriei* now grace California's streets, parks, and gardens. The hybrids, developed mostly at the U.S. National Arboretum in the 1960s, have the superior characteristics of both parent species, including the brilliantly colored pink, red, and purple summer flowers of *L. indica*, and the compact, single-trunked tree form, hardiness, and powdery mildew resistance of *L. fauriei*. Crape myrtles can be recognized by their dark green oval leaves, borne one or two per node on four-ridged stems, their cone-shaped clusters of frilly flowers, each with six thin, crinkled petals reminiscent of crepe paper, and their smooth, sinuous trunks patched with tan, cinnamon, and pink bark. These trees grow best in warm spots, in full sun, in well-drained, loamy soils, protected from coastal wind.

New leaves

Mature fruits

Variation
in flower
color

*If trees could scream, would we be so cavalier about
cutting them down? We might, if they screamed all
the time, for no good reason. —Jack Handey*

Lagunaria patersonia

Primrose Tree

lah-goo-NAIR-ee-ah pah-ter-SOW-nee-ah
Lagunaria - Andrés Laguna (1499–1559)
patersonia - William Paterson (1755–1810)
Malvaceae

Queensland, Norfolk Island,
Lord Howe Island
Simple, Alternate
Evergreen, 30–50 ft.

Many plants have evolved visual cues that encourage visitation by animals, which can be a doubled-edged sword, as animals can also eat and destroy them. Plants become attractive while flowering, when animals are needed to move pollen between individuals. Later, animals are discouraged from visiting by green, unripe fruits and then attracted once again, by colorful ripe fruits, when they are needed for seed dispersal. The primrose tree's floral buds are green, inconspicuous, and leaflike until they open for pollination, with spreading, velvety, magenta petals. Then, as if to say to animals, "Go away, you are not needed now," the unripe fruits mimic the green floral buds while seeds develop inside. Only when the seeds are mature do the fruits turn brown, split open, and reveal reddish seeds, beckoning birds for dispersal. The primrose tree, which is widely planted in Southern California, is a hibiscus relative and the only member of its genus. This species is occasionally referred to by the common name "cow-itch tree" because the sharp hairs in the open fruits can cause skin irritation.

Immature fruit with
developing seeds
and irritating hairs

Mature fruits and seeds

Shoot with leaves
and immature fruits

Trees conduct the eye from the ground up through the heavens; link the detailed temporaries of life with the bulging blue abstraction overhead. They alone seem to unite the earth and the sky—connecting the known world with everything that is beyond our grasp and our power. —Diane Ackerman

Laurus nobilis   Grecian Laurel or Sweet Bay

LORE-uss NOE-bih-liss
Laurus - L., laurel tree
nobilis - L., renowned
Lauraceae

Mediterranean
Simple, Alternate
Evergreen, 20–40 ft.

O f the 2,500 or so species in the aromatic, mostly tropical laurel family (Lauraceae), famous members include avocado (*Persea americana*), sassafras (*Sassafras albidum*), cinnamon (*Cinnamomum verum*), camphor (*C. camphora*), and sweet bay (*Laurus nobilis*). The fragrant leaves of sweet bay, which contain the essential oil cineole, have long been used in culinary seasoning. The more pungent leaves of the only native California member of the Lauraceae, the bay laurel (*Umbellularia californica*), are also occasionally used for flavoring. Sweet bay leaves were the classic Greek symbol of victory, honor, and heroics, worn in the form of crowns and garlands. This use is the basis of the modern phrase "rest on your laurels" and the titles "poet-laureate" and "baccalaureate"; the term for a bachelor's degree comes from the Latin *baccalaureus*, meaning "laurel berry." This slow-growing species is not particular about soil or water conditions and eventually becomes a large, multistemmed shrub or single-trunked tree. Hold one of its glossy leaves up to the light and the translucent margins will be clearly visible. The Saratoga bay laurel (*L.* 'Saratoga'), which has more egg-shaped leaves, is also widely grown.

Leaves with
translucent margins

Flowers

In all the vegetable world, trees have yet another attractiveness, much more mysterious and grand; because of their long lives, because their image is shaped by eternal forces, because their ascending verticality unites earth with sky, crossing the domain of humanity, trees seem the support most appropriate for all cosmic dreaming. —Francis Hallé

Ligustrum lucidum   Glossy Privet

lih-GUSS-trum LOO-sih-dum
Ligustrum - L., privet
lucidum - L., shining
Oleaceae

China, Japan, Korea
Simple, Opposite
Evergreen, 20–40 ft.

Flower clusters

Some trees are seemingly indestructible and grow anywhere. One of these is glossy privet, a ubiquitous background tree grown in all parts of California. It thrives in adverse soil conditions, drought, heavy wind, and temperature extremes. This tough species naturally forms a medium-sized, lollipop-shaped tree, but because it easily capitulates to the brutalities of pruning, it is often grown as a large shrub, hedge, or disfigured version of both. Glossy privet emerges from the background for a short time during the early summer when it blooms prodigiously with sweetly malodorous flowers. These flowers develop into spires of poisonous, bluish black fruits that fall from the tree staining everything below. The fruits are also dispersed readily by birds, which is why glossy privet has become a weed in regions of California.

Of all the wonders of nature, a tree in sum-
mer is perhaps the most remarkable; with
the possible exception of a moose singing
"Embraceable You" in spats. —Woody Allen

Polished-looking foliage
of *L. japonicum*

Fruits

Liquidambar styraciflua   Sweetgum

lih-kwid-AM-bar sty-RAH-sih-floo-uh

Luiquidus - L., fluid; *ambar* - Arabic, amber

styraciflua - L., flowing with resin

Altingiaceae

Connecticut to Central America

Simple, Alternate

Deciduous, 60 ft.

ew of California's commonly grown trees provide brilliant fall color, which explains the popularity of sweetgum. This species has an impressive native range, from Connecticut to Florida, west to Texas, and south to scattered populations in Mexico and Central America. The long-lasting fall foliage turns orange, yellow, crimson, burgundy, purple, and occasionally blue. Sweetgum can be recognized by its lobed, maple-like, aromatic leaves (crush one under your nose) and its spiny fruit clusters. Some individuals have corky, winged twigs. A fragrant, amber-colored fluid (styrax, similar to storax from oriental sweetgum) exudes from the inner bark. This bitter resin was once used as a medicinal chewing gum, in perfumery, and to flavor tobacco in Mexico. The wood is commercially logged for furniture and veneers.

Variation in fall leaf color

Without trees, a city is just a scab on the earth. —Chuck Gilstrap

Bark

Mature fruit

Corky stems

Liriodendron tulipifera

leer-ee-oh-DEN-dron too-lih-PIH-fer-uh
Leirion - Gr., lily; *dendron* - Gr., tree
tulipifera - L., tulip-bearing
Magnoliaceae

Eastern North America
Simple, Alternate
Deciduous, 80+ ft.

Tulip tree has the most exquisite, yet most inconspicuous flowers. They are tulip-like in shape but have six pale greenish yellow petals with bright orange blotches at the base. They often grow high in the tree and blend in with the canopy. A long pole trimmer, brave climb, jet pack, or view from a three-story building may be necessary to observe these majestic blooms closely, but they are well worth the effort. This magnolia relative is a widely esteemed native of the eastern U.S., where it grows two hundred feet tall or more—taller than any other tree in that region and taller than any other deciduous tree in the world! It is treasured for its straight, shapely form, distinctively truncated leaves (four-pointed with a notched tip), and lemon yellow to golden fall color.

Flower

The tree represents an inner world of the human soul; the unconscious realm whose vigor, spontaneity and fertility has been alienated by modern technological civilization. —Steven Marx

Old fruit cluster

Leaf

One-seeded, winged fruits

California's Largest Urban Trees

Bunya Bunya *(Araucaria bidwillii)*
Deodar Cedar *(Cedrus deodara)*
Camphor Tree *(Cinnamomum camphora)*
Blue Gum *(Eucalyptus globulus)*
Moreton Bay Fig *(Ficus macrophylla)*
Monterey Cypress *(Hesperocyparis macrocarpa)*
Tulip Tree *(Liriodendron tulipifera)*
Monterey Pine *(Pinus radiata)*
Coast Redwood *(Sequoia sempervirens)*
Elms *(Ulmus spp.)*

Lophostemon confertus   Brisbane Box

loh-foe-STEE-mon kun-FUR-tus

Lophos - Gr., crested; *stemon* - Gr., stamen

confertus - L., crowded

Myrtaceae

Synonym: *Tristania conferta*

Eastern Australia
Simple, Alternate
Evergreen, 40–60 ft.

Flowers with feathery bundles of stamens

Fruits

The attractive, low-maintenance, and reliable Brisbane box can easily be mistaken for other trees. Its smooth, luminous, rusty pink bark and dark green leaves are like those of the California native madrone (*Arbutus menziesii*). Madrone leaves, however, have distinctly lighter undersides. The upright, slender shape and woody capsules of the fast-growing Brisbane box make it difficult for the casual observer to distinguish between it and *Eucalyptus*. However, it lacks the defining characteristic of the closely related gum trees: petals fused into a cap that covers the unopened flowers. Also unlike eucalypts, Brisbane boxes bear stamens in bundles, and the leaves are unscented.

The great French Marshal Lyautey once asked his gardener to plant a tree. The gardener objected that the tree was slow growing and would not reach maturity for 100 years. The Marshal replied, "In that case, there is no time to lose; plant it this afternoon!" —John F. Kennedy

Lyonothamnus floribundus subsp. *aspleniifolius*   Island Ironwood

lye-on-owe-THAM-nuss floor-ih-BUN-duss
Lyono - William S. Lyon (1851–1916);
   thamnus - Gr., a shrub
floribundus - L., abundant flowers
aspleniifolius - L., leaves like spleenwort (*Asplenium*)
Rosaceae

Santa Barbara Channel Islands
Pinnate, Opposite
Evergreen, 50 ft.

sland ironwood is a beautiful evergreen tree that grows natively only on the Santa Barbara Channel Islands. There are two subspecies with different leaf shapes and native ranges. The only widespread and commonly grown subspecies, *aspleniifolius* (Santa Cruz Island ironwood), has pinnately compound leaves with scallop-toothed leaflets. The trunk's interesting bark peels and shreds with age, exposing smooth, cinnamon-colored young bark. Flat-topped clusters of white flowers are produced in spring, held above the foliage at branch tips, and can linger on the plant as dry, brown clusters for years.

The forest will answer you in the way you call to it.
—Finnish Proverb

Shoot with
opposite leaves

Magnolia grandiflora   Southern Magnolia

mag-NOE-lee-uh gran-dih-FLOR-ah

Magnolia - Pierre Magnol (1638–1715)

grandiflora - L., large-flowered

Magnoliaceae

Southeastern United States

Simple, Alternate

Evergreen, 40–80 ft.

Magnolias are an ancient and primitive group of flowering plants that evolved at a time when Earth was covered primarily with ferns and conifers. The iconic southern magnolia is widely planted throughout California and is the most widely grown ornamental tree on Earth. It can be recognized by the contrasting sides on its evergreen, stiff, leathery leaves: glossy, dark green above and gray to rust-colored and felted below. Its spectacular, fruit-scented, creamy white flowers are borne individually on the deep-green canopy like huge water lilies. They are California's largest cultivated tree flowers, some reaching a foot in diameter. These flowers evolved prior to butterflies and bees and were originally pollinated by beetles and other ancient insects. Pollinated flowers mature into aggregate cone-like clusters of small fruits, each splitting to unveil a fleshy, scarlet seed.

Seeds and mature fruits

Leaves

Few other trees bring such a whiff of the palace to suburbia like the southern magnolia.
—Thomas Pakenham

California's Ten Most Widely Cultivated Urban Trees (in order of prevalence)

Crape Myrtle *(Lagerstroemia* hybrids and culitvars*)*
London Plane Tree *(Platanus × hispanica)*
Mexican Fan Palm *(Washingtonia robusta)*
Sweet Gum *(Liquidambar styraciflua)*
Queen Palm *(Syagrus romanzoffiana)*
Chinese Pistache *(Pistacia chinensis)*
Southern Magnolia *(Magnolia grandiflora)*
Callery or Bradford Pear *(Pyrus calleryana)*
Coast Live Oak *(Quercus agrifolia)*
Canary Island Pine *(Pinus canariensis)*

Magnolia × *soulangeana*   Saucer Magnolia

mag-NOE-lee-uh soo-lan-jee-AH-nah

Magnolia - Pierre Magnol (1638–1715)

soulangeana - Étienne Soulange-Bodin (1774–1846)

Magnoliaceae

Hybrid

Simple, Alternate

Deciduous, 20–30 ft.

The most widespread of the many flowering magnolia hybrids, saucer magnolia is a small deciduous tree with magnificent flowers that appear before new leaves emerge. This hybrid was created in the garden of M. Soulange-Bodin outside Paris in the 1820s when two Asian species, yulan magnolia (*M. denudata*) and lily magnolia (*M. liliiflora*), were crossed. In midwinter, the ephemeral flowers—pink and purple on the outside, white on the inside—emerge from fuzzy buds that can be found among the light green leaves months before flowering.

Fuzzy
flower
bud

It is the presence of the tree
itself that moves us: its
spiritual qualities—qualities
of self-sufficiency, tenacity,
endurance, resilience, silence,
stability, egolessness—the
sheer quality of being—being
rather than doing—and in
many cases, of being alive
for hundreds of years.
—Steven Marx

Leaves

Malus spp.

MAL-uss

Malus - L., apple

Rosaceae

Hybrid

Simple, Alternate

Deciduous, 25 ft.

alus is a genus of about 40 species of large shrubs and small trees from the north temperates, particularly the Caucasus and Southwest Asia. The genus contains over two thousand named cultivars of the edible apple (*M. pumila,* syn. *M. communis* and *M.* × *domestica*), including the most widely grown variety, 'Red Delicious', as well as the popular 'Granny Smith', 'McIntosh', 'Fuji', and 'Golden Delicious', some of which originated as chance seedlings in crab apple orchards. Crab apple trees make small (ranging from the

Malus 'Donald Wyman' fruits

size of peas to golf balls), mostly inedible, often mealy apples, and are grown ornamentally for their flowers. For a brief time in mid-spring these subtly fragrant, white, pink, carmine, and even purplish flowers decorate front yards, gardens, and parks in much of California. Crab apples are some of the most common urban trees in America's colder cities, but within California they do not typically flower or grow well outside the coolest northern and inland municipalities.

Malus 'Liset' in full bloom

M. floribunda, red floral buds and white open flowers

The best friend on Earth of man is the tree: when we use the tree respectfully and economically we have one of the greatest resources of the earth.
—Frank Lloyd Wright

Maytenus boaria   Mayten Tree

may-TEH-nuss boe-AIR-ee-uh

Mantun - Mapuche Indian name

bovarius - L., of cattle

Celastraceae

Chile, Argentina

Simple, Alternate

Evergreen, 20–40 ft.

Bark

Mayten tree is one of the few ornamental trees in California native to Chile. In its native habitat, the small, toothed leaves, which are held perpendicular to the twigs, are popular as cattle fodder. They are also used medicinally, as an antiseptic and to reduce fever. This graceful tree can be recognized by its gray, checkered bark and evergreen crown of pendulous branches. The fruits split open, exposing seeds surrounded by a reddish orange, fleshy coat that attracts birds, aiding in seed dispersal.

Open fruit and seed

¼ in.

Flowers

I must confess that I have also transferred human feelings to trees. At the base of a great makoré tree, I cannot help but sense a sly haughtiness, as if the tree is amused by the strange new organism, bilateral and noisy, who suddenly lowers his pants to relieve himself, showing that he lives in a very different time scale, one in which the problems of waste removal have not been solved very elegantly.
—*Francis Hallé*

Melaleuca quinquenervia

meh-luh-LOO-kuh kwin-kweh-NUR-vee-uh

Melas - Gr., black; *leukos* - Gr., white

quinquenervia - L., five-nerved

Myrtaceae

Australia, New Guinea,
New Caledonia
Simple, Alternate
Evergreen, 20–40 ft.

ajeput tree is the most commonly culti-vated *Melaleuca* in California and possibly the world. Many species of *Melaleuca*, including the cajeput, grow in wetlands, salt marshes, and sand dunes in their native habitats. This species grows vigorously in poor, salty, and waterlogged soils, as well as in drought, heat, and wind. Although it is not invasive in California, its adaptability to harsh conditions is possibly why it has become a nasty weed in the Florida Everglades, quickly turning native wetlands into cajeput thickets. Cajeput bark is the source of a medicinal oil called niaouli.

Flowering shoot

Fruit clusters

California's Weediest Trees

Silver Wattle *(Acacia dealbata)*
Tree of Heaven *(Ailanthus altissima)*
Carrotwood *(Cupaniopsis anacardioides)*
Russian Olive *(Elaeagnus angustifolia)*
Blue Gum *(Eucalyptus globulus)*
Edible Fig *(Ficus carica)*
Myoporum *(Myoporum laetum)*
Black Locust *(Robinia pseudoacacia)*
Brazilian Pepper *(Schinus terebinthifolius)*
Chinese Tallow Tree *(Triadica sebifera)*

A man has made at least a start on discovering the meaning of human life when he plants shade trees under which he knows full well he will never sit. —Elton Trueblood

Someone's sitting in the shade today because someone planted a tree a long time ago. —Warren Buffett

Melaleuca spp. Other Paperbark Trees

All but a dozen or so of the 250 species of *Melaleuca* grow natively only in Australia. They tend to have thick, papery, spongy bark that darkens with age, then peels in sheets to reveal contrasting lightly colored bark beneath (hence the genus name's derivation from Latin words for "black" and "white"). Like *Callistemon*, *Melaleuca* has spikes of bottlebrush-like flowers and tightly packed woody fruits attached directly to branches. The widely grown flaxleaf paperbark (*M. linariifolia*) displays such profuse blossoms in early summer that it looks like bright white snow has fallen on it.

Flaxleaf Paperbark *(M. linariifolia)*

Flaxleaf Paperbark (*M. linariifolia*)

Prickly Paperbark (*M. styphelioides*)

Heath Melaleuca (*M. ericifolia*)

Pink Melaleuca (*M. nesophila*)

Drooping Melaleuca (*M. armillaris*)

Heath Melaleuca *(M. ericifolia)* bark

Key to Commonly Cultivated *Melaleuca*

1. Leaves >1.5 in. long—Cajeput Tree *(M. quinquenervia)*
1' Leaves <1 in. long
   2. Leaves oblong, oval, or lance-shaped
      3. Tree; flowers white; leaves sessile, sharply pointed—Prickly Paperbark *(M. styphelioides)*
      3' Large shrub; flowers pink; leaves with short petiole, abruptly pointed—Pink Melaleuca *(M. nesophila)*
   2' Leaves linear
      4. Leaves opposite or mostly so, especially on older growth—Flaxleaf Paperbark *(M. linariifolia)*
      4' Leaves alternate, scattered, or in whorls, especially on older growth
         5. Leaves usually >¾ in. long, tips with a distinct hook—Drooping Melaleuca *(M. armillaris)*
         5' Leaves usually <¾ in. long, tips straight or barely curving—Heath Melaleuca *(M. ericifolia)*

Melia azedarach   Chinaberry

MEL-ee-uh ah-ZEE-duh-rak
Melia - Gr., ash tree
azaddhirakt - Persian, ash tree
Meliaceae

China, North India to
northeastern Australia
Bipinnate, Alternate
Deciduous, 40–50 ft.

The genus *Melia* is the namesake of the Mahogany family (Meliaceae), whose members include some of the most important timber trees in the world: the true mahoganies (*Swietenia* spp.). Today, all species of mahogany, which have been used for centuries to make fine furniture, are listed by the Convention on International Trade in Endangered Species as threatened, and few wild stands remain uncut. Chinaberry has been planted for centuries for its mahogany-like wood, as a source of medicine and natural insecticide, and as an ornamental tree. Its fast growth and drought tolerance have made it a popular shade tree in many of the world's driest areas, including Arabia, the Mediterranean, and the drier interior parts of California. Chinaberry, also known as "bead tree," "pride of India," and "Persian lilac," can be recognized by its lacy, bipinnately compound leaves, its fragrant, purple-lilac flowers, and its golden tan, egg-shaped fruits, which are poisonous. Each houses a single, bony, corrugated pit; these are often polished for Asian rosary beads. Chinaberry is widely dispersed by birds and has become an invasive weed in many areas where it is cultivated, including the Central Valley and Southern California.

Cluster of ripe fruit

Fruit pits

A single, large,
bipinnately
compound leaf

Flowers

Trees do not seem to be aware, as dogs and monkeys are aware. They do not have brains. But they are sentient in their way; they gauge what's going on as much as they need to, and they conduct their affairs as adroitly as any military strategist. —Collin Tudge

Metrosideros excelsa — New Zealand Christmas Tree

meh-troe-SID-air-ose ek-SELL-suh
Metra - Gr., heartwood; *sideros* - Gr., iron
excelsa - L., tall
Myrtaceae

New Zealand's North Island
Simple, Opposite
Evergreen, 40–60 ft.

A erial roots, those arising from above-ground parts of a plant, are a common occurrence in the tropics. However, in California's low humidity, most trees that make aerial roots in their native habitats, such as the Moreton Bay fig (*Ficus macrophylla*), fail to do so here. Nevertheless, older New Zealand Christmas trees here often have well-developed aerial roots hanging from larger branches, particularly on trees lining foggy San Francisco streets. This species can be recognized by its opposite, leathery leaves, which are glossy, dark green above, densely hairy below, and slightly rolled under at the margins. Like many trees in the myrtle family (e.g., *Melaleuca, Callistemon, Eucalyptus*) the conspicuous and colorful parts of the flowers are not the petals but the numerous stamens. The New Zealand Christmas tree's scarlet stamens are crowned by golden anthers. The nectar-filled flowers are produced profusely in summer in both California and New Zealand. Since Christmas occurs during the Southern Hemisphere summer, the flowers are used as holiday decoration.

Fruits and flowers

*Between every two trees is
a doorway to a new world.*
—*John Muir*

Aerial roots

Michelia doltsopa

Sweet Michelia

mye-KELL-ee-uh dult-SOE-puh
Michelia - Pietro Antonio Mecheli (1679–1737)
doltsopa - Tibetan name for this tree
Magnoliaceae
Synonym: *Magnolia excelsa*

China, Bhutan, Nepal,
India, and Myanmar
Simple, Alternate
Evergreen, 30+ ft.

This broad-leaved evergreen tree, with its shiny, leathery leaves and velvety brown floral buds, could be mistaken for the southern magnolia (*Magnolia grandiflora*). However, the position of the flowers distinguishes the seventy or so michelias from the magnolias: whereas magnolias make flowers only at branch ends, the fragrant sweet michelia flowers, each a waxy, white saucer about four inches across, are borne along stems in the axils of leaves. Studies using DNA recently exposed this morphological difference as a poor indicator of evolutionary relatedness and some michelias have turned out to be more closely related to magnolias than other michelias. Although few michelias grow taller than 30 feet in California, they become magnificent timber trees in their native Himalayas, where the finely textured, durable, and easily worked wood is used for the construction of doors, window frames, and furniture. In California, sweet michelia usually grows best in milder, coastal weather. A hybrid between this species and the smaller banana shrub (*M. figo*) has a scent like imitation banana candy and is sold in California under the name *Michelia × foggii*.

Among all the varied productions with which Nature has adorned the surface of the earth, none awakens our sympathies, or interests our imagination so powerfully as those venerable trees which seem to have stood the lapse of ages, silent witnesses of the successive generations of man, to whose destiny they bear so touching a resemblance, alike in their budding, their prime and their decay. —John Muir

MORE-us AL-bah
Morus - L., mulberry tree
alba - L., white
Moraceae

China
Simple, Alternate
Deciduous, 20–50 ft.

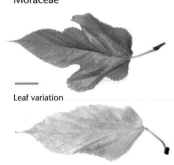

Leaf variation

The coarsely toothed, glossy leaves of the white mulberry, the primary food of silkworms (*Bombyx mori*), range in shape from generously lobed to completely unlobed, even on the same branch. It takes nearly ten thousand pounds of these leaves to make a single silk dress. Pull a leaf from a mulberry tree and you will find that it bleeds milky latex from the injury, a characteristic mulberries share with most other members of the Moraceae, such as *Ficus*. Inconspicuous, unisexual flowers are followed by fruits that look like small blackberries but taste like an insipid mixture of a raspberry and a fig. These fruits stain whatever they fall on, including sidewalks, clothing, and cars. Not surprisingly, the most popular mulberry varieties are fruitless, male trees with brilliant, yellow fall color. However, male mulberries are a major cause of hay fever. The black mulberry (*M. nigra*), which makes larger, tastier fruits, can be distinguished by its leaves, dull above and furry below, in contrast to the glossy, hairless leaves of the white mulberry.

The symbolism—and the substantive significance—of planting a tree has universal power in every culture and every society on Earth, and it is a way for individual men, women and children to participate in creating solutions for the environmental crisis. —Al Gore

Male flowers

Fruits

Myoporum laetum

mye-OPP-orr-um LAY-dum
Myo - Gr., to close; *poros* - Gr., opening
laetum - L., bright
Scrophulariaceae

New Zealand
Simple, Alternate
Evergreen, 30 ft.

Flowers

Myoporum (also known by its Maori name Ngaio) is well adapted to seaside climates and was once extensively planted throughout coastal parts of California as a dense screen tree and windbreak. It is considered a weedy and invasive tree by the California Invasive Plant Council, having fallen out of favor for a number of reasons: its aggressive surface roots may crack sidewalks and push up pavement; it has poisonous leaves, messy fruit, and an irregular and unruly growth habit; and it is severely damaged by frost. In addition, many of the myoporum trees in California have become infested with myoporum thrips (*Klambothrips myopori*) that distort leaves, form galls, and can eventually completely defoliate and kill mature plants. The scientific name for this tree alludes to a unique leaf character: *Myoporum* comes from two Greek words meaning "filled pore," referring to the translucent dots easily seen in the fleshy leaves when they are held to the light.

Myoporum thrips
infested shoot

When we have learned how to listen to trees, then the brevity and the quickness and the childlike hastiness of our thoughts achieve an incomparable joy. Whoever has learned how to listen to trees no longer wants to be a tree. He wants to be nothing except what he is. That is home. That is happiness.
—Herman Hesse

Olea europaea   Olive

OH-lay-ah yur-OH-pee-ah
Olea - L., olive
europaea - L., European
Oleaceae

Eastern Mediterranean
Simple, Opposite
Evergreen, 20–30 ft.

The iconic olive tree has enjoyed a long shared history with the human race; there is archaeological evidence of olive consumption nearly six thousand years ago. In this very long-lived species, some five-hundred-year-old trees still produce valuable crops. Olives were likely the first non-native tree cultivated in California, in 1769 at Mission San Diego. Ever since, they have been an important agricultural crop in the state, grown for table olives and olive oil, and also grown as shade, street, and park trees. Olives can be recognized by their gray, gnarled, picturesque trunks and by their slender, opposite, gray-green leaves with silver undersides. According to Italian folklore, drought, sun, rocky soil, silence, and solitude are the five main ingredients that create the ideal olive. Only the last two are lacking for olives on the streets of California's urban landscapes.

Flowers

Fruits

The olive tree is the richest gift of heaven. —Thomas Jefferson

Parkinsonia spp.

par-kin-SOWE-nee-uh
Parkinsonia - John Parkinson (1567–1650)
Fabaceae
Synonym: *Cercidium*

Arid parts of Africa and the Americas
Bipinnate, Alternate
Drought deciduous, 15–30 ft.

P alo verde trees are a group of about a dozen spiny, drought-deciduous, golden-flowered species native to desert watercourses of Africa and the Americas. Palo verde means green stick in Spanish, alluding to the fact that these picturesque small trees do much of their photosynthesis in their greenish-yellow bark while leafless. The finely-textured compound leaves shed their tiny leaflets during dry times of the year. Showy, fragrant yellow flowers with orange-spotted petals bloom in late spring and early summer. Several species are grown widely in California and elsewhere, and some have become naturalized weeds in the world's tropical and subtropical regions. A sterile, thornless hybrid called Desert Museum Palo Verde (*Parkinsonia* × 'Desert Museum') is an excellent, drought-tolerant small tree for inland gardens. It tends to suffer from powdery mildew infestation in moist marine air.

Desert Museum Palo Verde (*P.* × 'Desert Museum') flowers

The more you know about trees, the less time you get to spend with them. —Joe Bennassini

Jerusalem thorn (*P. aculeata*), spine

Desert Museum Palo Verde (*P.* × 'Desert Museum') leaf

leaflets

Palo brea (*P. praecox*) bark

Pistacia chinensis Chinese Pistache

pis-TAH-see-ah chi-NEN-sis
Pistake - Gr., nut
chinensis - L., Chinese
Anacardiaceae

China, Taiwan
Pinnate, Alternate
Deciduous, 40 ft.

The Chinese pistache is a member of a large and mostly tropical plant family. Other familiar members are the cashew (*Anacardium occidentale*), pistachio (*Pistacia vera*), mango (*Mangifera indica*), marula (*Sclerocarya birrea*), sumac (*Rhus* spp.), poison oak (*Toxicodendron diversilobum*), and poison ivy (*T. radicans*). Many members of this family make an allergenic toxin called urushiol that causes a skin rash on contact; however, concentrations are too low in Chinese pistache to cause contact dermatitis in most people. Instead, this reliable tree is a favorite in California landscapes as it is fast-growing and has a dense, rounded crown and beautiful orange and red fall foliage. Female trees bear small, spherical, bright red fruits that turn bluish black in the fall. The Chinese pistache is drought tolerant, can handle the constricted root space of a small sidewalk tree site, and does well as a garden tree in full sun, heat, and well-drained soil.

We know now that we are at the point of confrontation between machinery and vegetation, and so it is that the trees, the most conspicuous and valuable solace for the innate yearning for the green and growing, assume a new and crucial significance. —David G. Leach

Winter fruit color

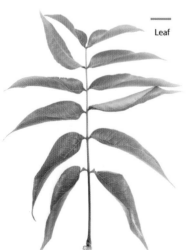

Leaf

Pittosporum undulatum   Mock Orange

pih-toe-SPORE-um un-doo-LAY-tum
Pitta - Gr., pitch; *spora* - Gr., seed
undulatum - L., wavy
Pittosporaceae

Eastern Australia
Simple, Alternate
Evergreen, 30–40 ft.

Mock orange's dense, rounded crown of bright green, polished, wavy-edged leaves makes it appealing as a street and park tree. However, it is the creamy, star-shaped flowers, exuding the sweet scent of orange and jasmine blossoms, that make it especially attractive. In the late afternoon of a warm California spring day, you can smell these trees from a great distance. The flowers develop into half-inch spherical orange fruits that split open to drop sticky black seeds, often staining the sidewalk below. Mock orange also goes by the names "Victorian box" and "cheese wood"— its cheesy-smelling wood has been used to manufacture golf club heads. In many tropical areas, including Hawaii, where this species is dispersed readily by birds, it is considered an invasive weed.

I willingly confess to so great a partiality for trees as tempts me to respect a man in exact proportion to his respect for them. —James Russell Lowell

Fruits

From the moving silence of trees, whether in storm or calm, in leaf and naked, night or day, we draw conclusions of our own, sustaining and unnoticed as our breath, and perilous also—though there has never been a critical tree— about the nature of things.
—Howard Nemerov

Pittosporum spp.
Other Pittosporums

T he nearly 200 species in the genus *Pittosporum* occur in tropical and subtropical regions of Australia, Africa, New Zealand, and Asia. Most of California's commonly cultivated pittosporums come from Australia and New Zealand and are grown primarily for their attractive, evergreen foliage. They are planted as street and park trees, garden shrubs, screens, and hedges. The name *Pittosporum* comes from the Greek words for pitch and seed; the seeds in this genus are embedded in a resinous, viscous fluid that helps them stick to birds and be dispersed over great distances. This attribute has helped pittosporum to become naturalized weeds in California and elsewhere.

Karo (*P. crassifolium*)

Lemonwood
(*P. eugenioides*)

Willow Pittosporum (*P. angustifolium*)

Kohuhu (*P. tenuifolium*)

Diamond Leaf Pittosporum
(*P. rhombifolium* syn. *Auranticarpa rhombifolia*)

Tobira (*P. tobira*)

Key to California's Commonly Cultivated *Pittosporum*

1. Leaf underside densely hairy—Karo (*P. crassifolium*)
1' Leaf underside hairless (glabrous)
   2. Leaves linear, ¼ in. wide or less—Willow Pittosporum (*P. angustifolium*)
   2' Leaves oval, spear-shaped, or diamond-shaped, ½ in. wide or wider
      3. Leaves coarsely toothed, diamond-shaped—Diamond Leaf
         Pittosporum (*P. rhombifolium* syn. *Auranticarpa rhombifolia*)
      3' Leaf margins toothless
         4. Leaf margins down-curved, not wavy—Tobira (*P. tobira*)
         4' Leaf margins wavy (undulate)
            5. Fruit diameter ⅜ in. or less, leaf mid-vein white—Lemonwood (*P. eugenioides*)
            5' Fruit diameter >½ in. leaf midvein green
               6. Leaves >4 in. long, fruit bright orange, flowers white—Mock Orange
                  (*P. undulatum*)
               6' Leaves <4 in. long, fruit greenish brown, flowers dark red, purple, or
                  black—Kohuhu (*P. tenuifolium*)

Karo (*P. crassifolium*)
flowers

Platanus spp.   Sycamores and Plane Trees

PLAT-tuh-nus

Platanos - Gr., broad

Platanaceae

Simple, Alternate

Deciduous, 50-100 ft.

S ycamores and plane trees are immense and picturesque, with muscular trunks and sculpted branches covered in mottled bark that sheds in patches to reveal pale, smooth bark below. They are iconic, long-lived trees in the wild and where they are grown. The large, often hairy, and sometimes rough-surfaced, palmately lobed leaves are subject to infection by the anthracnose fungus (*Apiognomonia veneta*), which causes wilting, dieback, and leaf drop. Some species and cultivated varieties are more resistant than others. The flowers are inconspicuous, wind-pollinated tight clusters that develop into dry seed balls dangling from the canopy. The genus has about eight species in North America, from Guatemala to California and eastern Canada, Europe through Western Asia, and Laos and Northern Vietnam. Although these species don't overlap in their native ranges, most form fertile hybrids when brought together in cultivation. Wild California sycamores (*P. racemosa*) growing near planted London plane trees (*P. × hispanica*) have been shown to hybridize in the Sacramento Valley area.

P. occidentalis

P. × hispanica

P. racemosa

P. mexicana   *P. orientalis* leaves   *P. racemosa* new growth

---

Key to California's Commonly Cultivated *Platanus*

1. Spherical fruit heads solitary—American Sycamore (*P. occidentalis*)
1' Spherical fruit heads 2 or more
   2. Leaves with 3 short triangular, shallow lobes; leaf undersides white hairy—Mexican Sycamore (*P. mexicana*)
   2' Leaves with 5 lance-shaped, long triangular, or elongate lobes; leaf underside hair or not
      3. Leaf lobe edges without teeth—California Sycamore (*P. racemosa*)
      3' Leaf lobe edges with prominent teeth
         4. Leaf lobes usually long, lance-shaped, deeper than ⅓ of the blade—Oriental Plane (*P. orientalis*)
         4' Leaf lobes usually triangular to lance shaped, shallower than ⅓ of the blade—London Plane Tree (*P. × hispanica*)

Platanus × *hispanica*

London Plane Tree

PLAT-tuh-nus hiss-PAN-ih-kah

Platanos - Gr., broad

hispanica - L., Spanish

Platanaceae

Synonym: *Platanus* × *acerifolia*

Hybrid

Simple, Alternate

Deciduous, 50–100 ft.

The London plane tree is the most commonly grown urban tree in the temperates. From Buenos Aires to New York City to Paris, Madrid, Sydney, and Shanghai, this hybrid of mysterious origin shades boulevards, parks, and squares. These stoic trees are resistant to air pollution, dust, root compaction, and other brutalities of urban living. Many authorities believe the London plane to be a hybrid between the Oriental plane (*P. orientalis*), from Eurasia, and the American sycamore (*P. occidentalis*), from the eastern United States. When these two species were brought into the confines of the same European garden (likely around Oxford in the late seventeenth century), vigorous hybrids resulted that quickly became popular in Europe. The London plane can be recognized by its leathery, woolly, maple-like leaves; its hanging, spherical clusters of flowers and fruit; and its mottled, olive-green to cream-colored bark. This tree is often subjected to a pruning practice called "pollarding" in which the new growth is cut off each year, resulting in a shapely, stunted tree with club-like branches.

Without exception, I've found that helping someone plant a tree is a positive experience, and people who plant trees are positive, forward-looking people.
—Jeffrey G. Meyer

Pollarded tree

Fruits

Platanus racemosa

California Sycamore

PLAT-tuh-nus race-MOE-suh
Platanos - Gr., broad
racemus - L., a bunch of grapes, for the floral clusters
Platanaceae

California
Simple, Alternate
Deciduous, 50–100 ft.

The fast-growing California sycamore can reach nearly one hundred feet in height in the right conditions. This species is found in canyons and near waterways in western California and Baja California. The velvety, palmately lobed leaves are shed during the winter, and spherical fruit clusters follow inconspicuous early spring flowers. The rigid older bark, which is incapable of expanding with the growing tree, flakes off in variable sections, exposing younger grayish-white patches. As bark is shed, the trunk and limbs take on their characteristic scaly, mottled look. The gnarled appearance of many of these trees is due to repeated infections by anthracnose, which can completely defoliate trees during the spring of wet years. Few trees die, however, because a second crop of leaves is produced in the early summer.

Trees go wandering forth in all directions with every wind, going and coming like ourselves, traveling with us around the sun two million miles a day, and through space heaven knows how fast and far! —John Muir

Flowers and new leaves

Bark of a young tree

Populus nigra 'Italica'   Lombardy Poplar

POP-yew-luss NYE-grah ih-TAL-lih-kah
Populus - L., poplar tree
nigra - L., black
Salicaceae

Western Asia, Europe
Simple, Alternate
Deciduous, 40–100 ft.

The Lombardy poplar historically has been planted along roadways, from country lanes in France to Pennsylvania Avenue in Washington, D.C., and alongside elevated freeways in San Francisco. It is easily recognized by its narrow, upright form, with branches hugging the main trunk and angling skyward to form a dense spire of shimmering foliage. It grows extraordinarily fast, sometimes reaching fifty feet in only fifteen years. A time-lapse video of the early life of a Lombardy popular would look something like a geyser of bright green, triangular, fluttering leaves hastily gushing upward. Unfortunately, individual trees seem to be short-lived, frequently succumbing to fungal diseases, usually from the top down, after forty or so years. Due to its propensity for sickness and its invasive roots, few nurseries still offer the Lombardy poplar for sale. All Lombardy poplars are clones that trace their origin to a single, mutant, male black poplar tree, possibly discovered in the seventeenth century somewhere in the Lombardy region of northern Italy, but maybe much earlier in Asia. Because Lombardy poplars are all males and produce no seeds, their reproduction is limited to asexual (vegetative) means—primarily by cuttings or root suckers. Might this mean that every Lombardy poplar is as old as the initial discovery some four hundred years ago? Are not all clonally propagated plants, such as bananas and potatoes, potentially immortal?

Fall leaves

A tree says: My strength is trust. I know nothing about my fathers, I know nothing about the thousand children that every year spring out of me. I live out the secret of my seed to the very end, and I care for nothing else. I trust that God is in me. I trust that my labor is holy. Out of this trust I live.
—Herman Hesse

Angiosperm: Eudicot: Malpighiales : **133**

PROO-nus sair-ah-SIH-fer-ah
Prunus - L., plum or cherry
cerasifera - L., cherry-bearing
Rosaceae

Balkans to Central Asia
Simple, Alternate
Deciduous, 30 ft.

Blireiana plum (*P.* × *blireiana*)

Fruit

P. cerasifera 'Atropurpurea'
with purple-bronze leaves

The many bronze- and purple-leaved cultivars and hybrids of *P. cerasifera* are widely grown in California. This species is also known as "Pissard plum," named after a gardener to the shah of Iran who supposedly introduced the tree into Western cultivation in 1880. The many cultivars, particularly 'Thundercloud', are more numerous than any other type of tree on the streets of San Francisco. In late winter they burst into bloom with light pink to white flowers that wither as new coppery red leaves emerge. In early summer, when the leaves have intensified to dark purple, the slightly sour yet tasty little plums are ripe and ready for eating. The fragrant, double-flowered Blireiana plum (*P.* × *blireiana*) is a popular hybrid of the purple leaf plum (*P. cerasifera* 'Pissardii') and the Japanese apricot (*P. mume* 'Alphandii').

What is more mortifying than to feel that you have missed the plum for want of courage to shake the tree?
—Logan Pearsall Smith

P. cerasifera 'Thundercloud' cultivar (the most common in California) in full bloom

Prunus spp.

Flowering Cherries and Cherry Laurels

PROO-nus

Prunus - L., plum or cherry

Rosaceae

North Temperates

Simple, Alternate

Deciduous or Evergreen, 20–30 ft.

The economically valuable genus *Prunus* contains many delectable stone fruits (cherries, peaches, plums, apricots, and nectarines). The 200 or so species also include both evergreen and deciduous ornamental trees and large shrubs, prized throughout California for their impressive winter and early spring floral displays, fine foliage, and shapely figures.

The rosy-pink, double-flowered Kwanzan cherry (*P. serrulata* 'Kwanzan' or *P. speciosa* 'Kwanzan') is one of many ornamental Japanese flowering cherry hybrids and cultivars collectively known in Japan as *sato zakura*.

The densely branched Taiwan cherry (*P. campanulata*), with its striking, reddish pink, bell-shaped, drooping flowers, grows well in warmer parts of California.

Carolina Laurel (*P. caroliniana*)

Key to California's Commonly Cultivated *Prunus*

1. Trees evergreen; leaves leathery, dark green; flowers white, borne in spikes
    2. Leaves <2 in. long—Hollyleaf Cherry (*P. ilicifolia*)
    2' Leaves >2 in. long
        3. Leaf margins smooth—Carolina Laurel (*P. caroliniana*)
        3' Leaf margins finely and evenly sawtoothed or with a few spiny teeth, mostly toward the tip
            4. Leaf tips narrowing to a point—Portugal Laurel (*P. lusitanica*)
            4' Leaf tips rounded and abruptly short-pointed—Cherry Laurel (*P. laurocerasus*)
1' Trees deciduous; leaves thin, bright green, bronze, or purple; flowers white to red, not borne in spikes
    5. Leaves usually purple or bronze when mature; tree blooming late winter—Purple Leaf Plum (*P. cerasifera*)
    5' Leaves green when mature; tree blooming late winter or spring
        6. Flowers red to dark pink, pendulous, petals bell-shaped—Taiwan Cherry (*P. campanulata*)
        6' Flowers white to pink, upright or pendulous, petals spreading
            7. Trunk warty, rough; flowers scented—Blireiana Plum (*P.* × *blireiana*)
            7' Trunk smooth (below graft union); flowers mostly scentless—Japanese Flowering Cherries (*P. serrulata* or *P. speciosa*)

Kwanzan cherry (*P. serrulata* 'Kwanzan') trunk

Pyrus calleryana Callery or Bradford Pear

PYE-rus kal-er-ee-AWN-ah
Pyrus - L., pear
calleryana - Joseph Callery (1810–1862)
Rosaceae
Synonym: *Pyrus taiwanensis*

China and Japan
Simple, Alternate
Deciduous, 20–50 ft.

Fall leaf color

Fruits

If I cherish trees beyond all personal (and perhaps rather peculiar) need and liking for them, it is because of this, their natural correspondence with the greener, more mysterious processes of mind—and because they also seem to me the best, most revealing messengers to us from all nature, the nearest its heart.
—John Fowles

During the last half of the twentieth century, Bradford pear was so favored by landscape architects and urban planners that it is now one of America's most ubiquitous street trees. The purpose of the initial introduction of *P. calleryana* to the U.S. from China in 1918 was agricultural, to breed it with the fruiting pear (*P. communis*) in hopes of generating fireblight-resistant orchards. Although that goal was never realized, one serendipitous outcome of the breeding program was the USDA's release in 1963 of an ornamental cultivar of *P. calleryana* called "Bradford," named for the USDA's Plant Introduction Station chief, Frederick Bradford. Bradford pear was publicized as the perfect street tree: fast-growing and disease resistant, tolerant of urban life, with a handsome pyramidal shape, brilliant fall color, and copious spring blossoms. Its popularity has led to its being found on the streets of many U.S. cities in higher numbers than any other species. Over time, however, a major weakness of the Bradford pear has become apparent: large, brittle, secondary branches grow crowded on the trunk at narrow vertical angles, making the tree vulnerable to self destruction by limb breakage. This fatal flaw is exacerbated in areas with high winds and has resulted in a disparaging backlash against the Bradford pear: many cities have removed it from their accepted street tree lists, and other cultivars are now more commonly planted.

Pyrus kawakamii   Evergreen Pear

PYE-rus kah-wuh-KAHM-ee-eye   Taiwan
Pyrus - L., pear   Simple, Alternate
kawakamii - Takiya Kawakami (1871–1915)   Evergreen, 20–30 ft.
Rosaceae
Synonym: *Pyrus calleryana*

Evergreen pear is only semi-evergreen in most of California. Depending on the local climate, some—occasionally all—of the brittle, reddish brown bodies of withered leaves will fall briskly to the ground and scrape along the sidewalk with the late-fall wind. If not pruned carefully, this tree often develops an erratic form, with branches like scaly bent knees, angling and jutting in various directions. This species can be spectacular in midwinter, when it is fleetingly covered with bright white, subtly fragrant flowers. When not in bloom, it can be recognized by its glossy, oval leaves with rounded teeth on the margins, and its charcoal gray bark that cracks into irregular squares with age.

Typical bark

Let us try to understand what trees are and we will be perplexed by their mixture of unchanging presence yet complete otherness. —Francis Hallé

Quercus spp.

KWER-kus

Quercus - L., name for these trees

Fagaceae

The world's 400 or so species of oaks are distributed in north temperates and higher elevations in the tropics. In Europe and North America, these trees are large, long-lived, and revered as symbols of strength. They have an extensive history of human use: many species make highly prized and strong wood used in furniture manufacturing, ship building, house construction, flooring, and cooperage for wines and spirits. Tannins and dyes are harvested from oak bark, and commercial cork comes from the spongy bark of the cork oak (*Q. suber*). Acorns are fed to animals and were the most important food source for Native Californians. Many of California's 20 or so native species and many commonly grown non-native oaks are important ornamental trees in urban areas.

Q. virginiana

Q. agrifolia

Q. rubra

Trees are the best monuments that a man can erect to his own memory. They speak his praises without flattery, and they are blessings to children yet unborn. —Lord Orrery

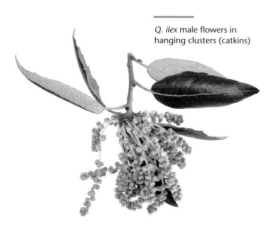

Q. ilex male flowers in hanging clusters (catkins)

North Temperates,
Upper Elevation Tropics
Simple, Alternate
Evergreen or Deciduous, 30–100 ft.

Oaks make small, wind-pollinated male and female flowers in separate locations on the same tree: the pollen-bearing male flowers hang in pendulous clusters, and the solitary, inconspicuous female flowers are borne in leaf axils. The female flower develops into a single-seeded nut (the acorn) that is partially enclosed in a cuplike cap made of concentric rings of tiny, overlapping scales. Identification of oaks can often be difficult since great variation exists within species and hybridization between species is common. Several native California oaks are widely grown as landscape plants, especially coast live oak (*Q. agrifolia*) and valley oak (*Q. lobata*), as well as non-native oaks, especially holly oak (*Q. ilex*) and cork oak (*Q. suber*). In addition, several deciduous oaks, primarily from the eastern U.S., are frequently cultivated in California's cooler interior areas for their stateliness and brilliant fall color.

Q. engelmannii bark

Q. douglasii trunk

*The creation of a thousand
forests is in one acorn.*
—*Ralph Waldo Emerson*

Q. lobata

Key to California's Commonly Cultivated and Native Oaks

1. Leaves evergreen, thick, leathery; leaf edges usually toothed or smooth, but usually not lobed
    2. Leaf underside shiny, hairless, green or greenish yellow, occasionally with small tufts of hair in vein axils
        3. Leaves convex, undersides with small tufts of hair in vein axils—Coast Live Oak (*Q. agrifolia*)
        3′ Leaves flat or only slightly convex, undersides hairless
            4. Leaves rarely longer than 3 in., usually 2 in. or shorter; leaf underside usually shiny, yellowish green—Interior Live Oak (*Q. wislizeni*)
            4′ Leaves up to 5 in. long, usually 1.5 in. to 3.5 in. long; leaf underside usually dull green— Shreve Oak (*Q. parvula* var. *shrevei*)
    2′ Leaf underside dull, white, or felt-like with golden or gray hairs (hairs sometimes vanishing as leaves age)
        5. Bark of trunk and large branches thick, soft, and corky—Cork Oak (*Q. suber*)
        5′ Bark of trunk and large branches thin, hard, or scaly, not soft and corky
            6. Lateral leaf veins straight and more or less parallel; acorn nut shell more or less woolly inside; acorns maturing in 2 years
                7. Lateral leaf veins prominently grooved into upper surface, raised above lower surface; young leaves not covered with golden hairs, hairs persistent on lower surface; acorn cup slightly wider than acorn body, without golden hairs—Island Oak (*Q. tomentella*)
                7′ Lateral leaf veins not generally grooved into upper surface; young leaves covered with golden hairs that fall off as leaves mature; acorn cup much wider than acorn body, covered with golden hairs—Canyon Live Oak (*Q. chrysolepis*)
            6′ Lateral leaf veins curved and usually not parallel; acorn nut shell hairless inside; acorns maturing in 1 year
                8. Leaf upper surface bluish green, dull—Engelmann Oak (*Q. engelmannii*)
                8′ Leaf upper surface dark green, shiny
                    9. Leaf tip tapered to a point; leaf blade usually widest at the middle—Holly Oak (*Q. ilex*)
                    9′ Leaf tip usually rounded, though sometimes tapered to a point; leaf blade usually widest distally—Southern Live Oak (*Q. virginiana*)
1′ Leaves deciduous, thin, supple; leaf edges usually distinctly lobed or wavy, sometimes smooth or toothed in *Q. douglasii* and *Q. engelmannii*
    10. Leaf lobe tips spiny, pointed, or bristly; acorn cup scales thin, papery; mature bark smooth or deeply furrowed, rarely scaly (sometimes scaly in *Q. kelloggii*)
        11. Acorn cup deep, usually covering ½ or more of the acorn; acorn cup scale tips loose; fall leaf color yellow to orange, never red—Black Oak (*Q. kelloggii*)
        11′ Acorn cup shallow, usually covering less than ½ of the acorn; acorn cup scale tips pressed tightly together; fall leaf color often red or reddish brown
            12. Leaf lobes usually extending less than half the distance to the midrib and tapering to their tips—Red Oak (*Q. rubra*)
            12′ Leaf lobes cut more than halfway to the midrib and not tapering evenly to their tips
                13. Lower branches drooping; leaf underside with conspicuous tufts of hair in vein axils; fall leaf color orange, brown—Pin Oak *(Q. palustris)*
                13′ Lower branches erect; leaf underside with obscure tufts of hair in vein axils; fall leaf color bright red—Scarlet Oak (*Q. coccinea*)
    10′ Leaf lobe tips generally rounded, not spiny or bristly; acorn cup scales thick, warty; mature bark scaly
        14. Leaf margins smooth, toothed, or shallowly lobed, sinuses between lobes less than ½ in.;upper surface of mature leaf bluish green or gray-green
            15. Leaves thin, falling in winter—Blue Oak (*Q. douglasii*)
            15′ Leaves somewhat leathery and often retained on twigs until spring—Engelmann Oak (*Q. engelmannii*)
        14′ Leaf margins deeply lobed, sinuses between lobes more than ½ in.; upper surface of mature leaf dark green
            16. Leaves mostly more than 4 in. long
                17. Leaf bases heart-shaped, with two lobes; acorn cup without soft awns along rim— English Oak (*Q. robur*)
                17′ Leaf base tapering to stalk; acorn cup with soft awns on scales forming fringe along rim—Bur Oak (*Q. macrocarpa*)
            16′ Leaves mostly less than 4 in. long
                18. Leaf lobes rarely overlapping; acorn cup usually more than ¼ in.; acorns usually more than 1 in. long, tip pointed—Valley Oak (*Q. lobata*)
                18′ Leaf lobes of some leaves touching or overlapping; acorn cup usually ¼ in. deep or less; acorns usually less than 1 in. long, tip rounded—Oregon Oak (*Q. garryana*)

Q. agrifolia Q. lobata Q. palustris Q. rubra Q. coccinea Q. robur

Q. rubra acorns

Q. tomentella acorns

Q. engelmannii acorn

Q. hypoleucoides

Q. garryana acorns

Q. macrocarpa acorns

Q. virginiana

Q. chrysolepis

The acorn does not know that it will become a sapling. The sapling does not remember when it was an acorn, and only dimly senses that it will become a mighty oak. The oak recalls fondly when it was a sapling, loves being a mighty oak, and joyfully creates new acorns. —James Earp

Quercus agrifolia   Coast Live Oak

KWER-kus ag-rih-FOE-lee-uh
Quercus - L., name for these trees
agrifolia - L., rough leaves
Fagaceae

California
Simple, Alternate
Evergreen, 40–70 ft.

Leaf underside showing tufts of hairs

Acorns

oast live oak grows natively in a variety of habitats from northern Sonoma County southward to northern Baja California. It is the dominant tree of coastal oak woodlands, a revered California icon, and the basis for city names such as Thousand Oaks and Oakland. Coast live oak is widely cultivated for its dense foliage, drought tolerance, and shapely beauty. Spanish settlers associated this tree, which they called *encina,* with fertile lands, and the locations of the Franciscan missions closely match the coast live oak's native range. This species can be distinguished from other native, evergreen oaks by its cupped leaves that bear scattered, spiny, marginal teeth and diagnostic tufts of fuzzy hairs in the axils of the main vein and secondary veins on leaf undersides.

Trees outstrip most people in the extent and depth of their work for the public good. —Sara Ebenreck

Trunk

Quercus ilex   Holly Oak

KWER-kus EYE-leks
Quercus - L., name for oak trees
ilex - L., name for this tree
Fagaceae

Mediterranean
Simple, Alternate
Evergreen, 50–70 ft.

The holly oak, also known as the holm oak, is California's most widely grown non-native oak. In urban settings it is often preferred over the similar-looking coast live oak (*Q. agrifolia*), due to its less invasive roots and more symmetrical, upright crown. The holly oak has been cultivated and revered for thousands of years in its native range surrounding the Mediterranean Sea. In Virgil's first-century epic poem *The Aeneid*, the Roman hero Aeneas picked the "golden bough" that would guide him on his journey through the underworld from a holly oak. *Ilex* is the ancient Roman name for this tree, a designation now used as the genus name for all hollies. The holly oak can be recognized by its scaly, gray bark and somber crown of dark green, leathery leaves (some looking like holly leaves), each with woolly, white or tan undersides.

Leaves and acorn, showing pointed leaf tips and white undersides

Bark

Pendent cluster of male flowers

Acorns

I part the out thrusting branches and come in beneath the blessed and the blessing trees. Though I am silent there is singing around me. Though I am dark there is vision around me. Though I am heavy there is flight around me. —Wendell Berry

Quercus lobata   Valley Oak

KWER-kus low-BAH-tuh

Quercus - L., name for these trees

lobata - L., lobed

Fagaceae

California

Simple, Alternate

Deciduous, 100+ ft.

alley oaks, which are the largest oaks in California, grow natively only in the Central Valley, Coast Ranges, and Sierra Nevada foothills, from Shasta County southward to Los Angeles County. These majestic trees can be recognized by their thick trunks covered with deeply checkered, alligator-skin bark and their broad, rounded canopies of craggy branches and deeply lobed leaves. The valley oak's sizable acorns, which can be up to two and a half inches long, were—along with many other types—the daily food source for most tribes of Native Californians. Valley oaks thrive in fertile soils with a high water table and were therefore extensively cleared for agriculture over the last hundred and fifty years. Many of the state's agricultural valleys, once home to vast stands of valley oak, now harbor meager numbers of enormous, remnant individuals.

Leaf

Trees veil many a house or building from critical scrutiny; they are the most therapeutic antidote to a city's architectural problems. —Palo Alto Chamber of Commerce, Trees of Palo Alto

Acorns

Suburbia is where the developer bulldozes out the trees, then names the streets after them.
—Bill Vaughn

Bark

KWER-kus SOO-ber

Quercus - L., name for oak trees

suber - L., corky

Fagaceae

Southern Europe, Northern Africa

Simple, Alternate

Evergreen, 70–100 ft.

C ommercial cork comes from the thick, spongy bark of the cork oak tree. Where it is harvested, the outer bark of each tree is skillfully and harmlessly stripped off the trunk and lower limbs once a decade, allowing new bark to regrow. The bark from some 250,000 trees is harvested annually, half of them in Portugal. In California, where cork is rarely harvested, these oaks, with their spacious, olive green crowns and stately trunks, are grown as ornamental trees. Next time you see a cork oak, take a moment to press your hand against the trunk. As the waxy, elastic bark yields beneath your touch, speculate why such a material would have evolved in the dry, fire-prone woodlands surrounding the western Mediterranean.

Trunk

Piece of
harvested bark
with removed
wine corks

Acorn and acorn cups
with overlapping scales

Leaves

Trees help you see slices of sky between branches, point to things you could never reach.... Trees take the eye to where it is, where it was, then over to distant hills, faraway to other places and times, long ago. A tree is a lens, a viewfinder, a window. I wait below for a message of what is yet to come.
—*Rochelle Mass*

Quercus virginiana   Southern Live Oak

KWER-kus ver-jin-ee-AH-nuh

Quercus - L., name for these trees

virginiana - L., for Virginia

Fagaceae

Southeastern U.S. and northern Mexico

Simple, Alternate

Evergreen, 40–60 ft.

Leaves showing variation in size and shape

When I place my own strong brown hand on the trunk of a tree, I feel connected to something that deserves my curiosity, care, and protection.
—Nalini Nadkarni

Leaves

Bark

The southern live oak, which is widely grown in California, is an iconic tree native to the southeastern U.S. coastal plain forests from Virginia to Florida and west to Texas, Oklahoma, and Mexico. Eventually, this species becomes massive and can live for hundreds of years. It's recognizable by its toothless leaves that are usually wider at the tip, shiny on the upper surface, and white or bluish underneath. The southern live oak evolved in rich soils with summer rain and thrives where it gets regular water. This species was an important food source for Native Americans and early European settlers, as the swollen, underground stems of seedlings were collected and eaten like a starchy potato. The dense wood was also highly valued for ship parts, and old-growth southern live oak groves were nearly all harvested by the late nineteenth century.

Robinia pseudoacacia   Black Locust

roe-BIN-ee-ah soo-doe-uh-KAY-shah
Robinia - Jean Robin (1550–1629)
pseudos - Gr., false; *akakia* - Gr., acacia
Fabaceae

Eastern North America
Pinnate, Alternate
Deciduous, 40–50 ft.

The roots of the black locust, like those of many other legumes, enjoy a symbiotic relationship with nitrogen-fixing bacteria, which convert nitrogen gas from the atmosphere to usable fertilizer for the tree. This relationship helps the black locust grow vigorously in poor soils. In its native Appalachian and Ozark habitats it rapidly colonizes disturbed areas, and it is often weedy in California, especially near abandoned home sites and along roadsides and watercourses. The black locust is a popular ornamental tree in Europe, where it was first introduced to France by Jean Robin, gardener to King Henry IV, in the early seventeenth century; some black locusts in and around Paris are over three hundred years old. It is still one of the most popular plantation trees in Europe. Its hard, decay-resistant wood is used for fence posts, veneers, cabinetry, and firewood. The black locust can be recognized by its pendent clusters of white, edible, sweetly fragrant, wisteria-like flowers, two sharp spines at the base of each leaf, and deeply furrowed, stringy bark. The ornamental hybrid *Robinia* × *ambigua* has pale pink to bright reddish pink flowers and is widely planted as a smaller street and garden tree throughout California.

Stipular spines

R. × *ambigua* 'Idaho' flowers

R. pseudoacacia flowers

Fruits

Robinia × ambigua

R. pseudoacacia bark

Schinus molle   Peruvian Pepper Tree

SHY-nus MOLL-ee

Schinos - Gr., mastic tree

molle - Quechua name for the tree

Anacardiaceae

Southern Andes
Pinnate, Alternate
Evergreen, 30–40 ft.

S. molle fruits and leaves

S. terebinthifolius fruits and leaves

Peruvian pepper was first planted in California by Father Antonio Peyri in the early 1800s at Mission San Luis Rey de Francia in Oceanside, and since then it has been a popular ornamental. Having been grown in California for so long, it is often misleadingly called the California pepper tree. This species is tolerant of extreme heat, drought, and virtually any soil conditions. It can be recognized by its gnarled trunk, spreading crown of pendulous, aromatic foliage, clusters of cream-colored flowers, and spherical rose-pink fruits. These spicy fruits, which are toxic in large amounts, are occasionally sold as pink peppercorns or mixed with true pepper (*Piper nigrum*). Brazilian pepper tree (*Schinus terebinthifolius*) is also widely grown in California and can be distinguished by its larger, oval leaflets. Both species, whose fruits are dispersed by birds, are weedy in many parts of the world, including Southern California.

People in suburbia see trees differently than foresters do. They cherish every one. It is useless to speak of the probability that a certain tree will die when the tree is in someone's backyard.... You are talking about a personal asset, a friend, a monument, not about board feet of lumber. —Roger Swain

Searsia lancea   African Sumac

seer-SEE-uh LAN-see-ah
Searsia - Paul B. Sears (1891–1990)
lancea - L., lance-shaped
Anacardiaceae
Synonym: *Rhus lancea*

South Africa, Namibia, Zimbabwe
Palmate, Alternate
Evergreen, 20–30 ft.

Bark

Fruits

*S*earsia is a genus of resinous trees and shrubs from warmer temperate parts of the world. African sumac is from dry savannas, *Acacia* woodlands, and along rivers and streams in southern Africa, where it is known as *karee*. It can be recognized by its dark green leaves, each divided into three linear leaflets that have a turpentine smell when crushed. The leaf stalk is conspicuously grooved on the upper surface. African sumac is popular for its drought tolerance, its graceful, weeping crown of russet red branchlets, and its gnarled trunk covered with scaly bark. In early winter it produces tiny, subtly fragrant, yellow flowers that are followed by pea-sized fruits on female trees. Unfortunately, this species is weedy, reproducing prolifically on its own in wild areas, especially in the desert washes and along watercourses in southeastern California.

Without trees our species would not have come into being at all. And if trees had disappeared after we hit the ground, we would still be scrabbling like baboons (assuming the baboons had even allowed us to live).—Collin Tudge

Male
inflorescence

Spathodea campanulata   African Tulip Tree

spath-OWE-dee-uh kam-pan-yoo-LAH-tuh

Spathe - Gr., spathe or boat-like calyx;

   odes - Gr., looks like

campanulata - L., bell-shaped

Bignoniaceae

Tropical Africa

Pinnate, Opposite

Evergreen, 30–50 ft.

The African tulip tree is the only species in its genus and a world-famous flowering tree. It's a popular street, park, and garden tree in the frost-free parts of coastal Southern California from Santa Barbara to San Diego. It is grown in tropical countries worldwide and has become invasive in many areas, including Hawaii, where birds readily pollinate the flowers, and where the winged seeds disperse over great distances. California is too dry for widespread natural reproduction of many cultivated trees that have become invasive in other places with higher rainfall. Examples of species that are weedy elsewhere but not in California are the jacaranda (*Jacaranda mimosifolia*, page 102), Brazilian pepper tree (*Schinus terebinthifolius*, page 148), cajeput tree (*Melaleuca quinquenervia*, page 118), and African tulip tree. In late summer and fall, when the African tulip tree starts to produce its spectacular blooms, the large unopened flower buds, which are filled with watery nectar, can be squeezed to make a botanical water pistol.

Yellowish-orange cultivar

Flowers

Large pinnately compound leaf

The tree of the field is man's life.
—Deuteronomy 20:19

Stenocarpus sinuatus

Firewheel Tree

sten-owe-KAR-puss sin-yoo-AYE-tuss
Stenos - Gr., narrow; *karpos* - Gr., fruit
sinuatus - L., wavy-margined
Proteaceae

Eastern Australia
Simple, Alternate
Evergreen, 30 ft.

Fruit with winged seed

The distinctive firewheel tree is a member of the eclectic Southern Hemisphere Protea family. Its flower clusters have a sensational arrangement in which individual, fire-red flowers radiate from a central point, like the spokes of a wheel. As each flower matures, the slender spokes turn from green to yellow, then crimson, with a globular, golden yellow ball at their apex, so that they resemble unlit matches. Glossy, dark green leaves range in size and shape from over a foot long and deeply lobed to less than four inches long and completely unlobed. The firewheel tree is slow-growing, often taking many years to its first bloom, then flowering intermittently throughout the year, eventually blooming most profusely during the summer heat. The flowers have a pungent, fetid smell, most apparent in the evening, which likely attracts moth pollinators. The scientific name of the genus alludes to the flat, narrow follicles that split open when mature, releasing winged seeds.

Leaf shape and size diversity

Developmental series of the flower cluster

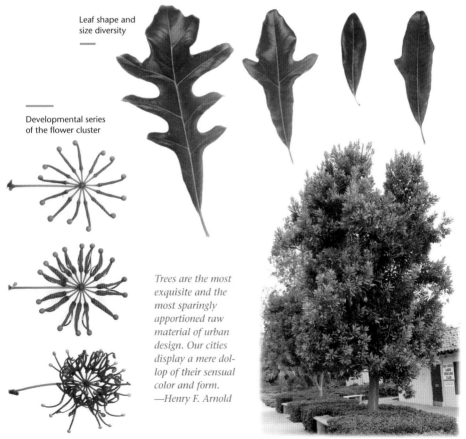

Trees are the most exquisite and the most sparingly apportioned raw material of urban design. Our cities display a mere dollop of their sensual color and form.
—Henry F. Arnold

Styphnolobium japonicum   Japanese Pagoda Tree

styfe-no-low-BEE-um juh-PON-ih-kum

Styphno - Gr., sour, astringent; *lobion* - Gr., pod

japonicum - from Japan

Fabaceae

Synonym: *Sophora japonica*

♂
♀

China, Korea
Pinnate, Alternate
Deciduous, 40–60 ft.

apanese pagoda tree, which is actually not native to Japan, has been cultivated there for thousands of years. It derives its common name from the fact that it is often planted in Asian Buddhist temple grounds. This tree is popular in many of the world's temperate cities for its pea-like, cream-colored, summer flowers and tolerance of urban conditions (it is the most abundant street tree in Madrid for instance). It is cultivated sporadically throughout much of California, predominantly as a lawn tree, but it is not popular as a street tree here, possibly because its fleshy legumes stain sidewalks. These fruits feel slimy and soapy to the touch when opened and are filled with poisonous seeds.

Fruits

Leaf

The works of a person begin imme-
diately to decay, while those of him
who plants begin directly to improve.
—William Shenstone

Flowers

sih-ZIH-jee-um oss-TRAW-lee

Syn - Gr., together; *zygon* - Gr., yoke

australe - L., southern

Myrtaceae

Eastern Australia

Simple, Opposite

Evergreen, 30–60 ft.

O f the 1,000 species in the large genus *Syzygium*, the brush cherry is the only widely grown member in California. It is often trimmed into a large hedge, but when allowed, it grows into a broadly columnar tree. It is prized for its appealing foliage (often with bronze-red new growth), creamy white summer blooms, and rose-purple fruits. This species is often erroneously sold under the name *S. paniculatum*, which is a rare species, also from eastern Australia and only occasionally found in California. *Syzygium australe* is distinguished by having four-angled, squarish

Fruits

stems with wings that join in pairs above each node, producing two small protrusions. The aromatic dried flower buds of the related species *S. aromaticum*, also known as cloves, are used as a spice throughout the world.

Opened fruit with a single multi-embryonic seed inside

Flowers

Four-angled stems with two protrusions near each node

Tamarix aphylla   Athel Tree

TAM-uh-ricks aye-FILL-ah

Tamarix - L., for the Tambre River,
   in northwestern Spain

a - Gr., without; *phyllon* - Gr., leaf

Tamaricaceae

North Africa, Europe, and Western Asia

Scale-like, Alternate

Evergreen, 30–50 ft.

amarix is a genus of about sixty shrubs and trees, several of which were introduced to California as landscape ornamentals, then have since become invasive. The athel tree, the largest member of the genus, is planted in California's deserts as a wind and sand break along roads and train tracks. The tiny, scale-like leaves excrete salt, often becoming encrusted in a thin white layer. The athel tree is remarkably useful, growing where few others will—in hot, dry climates; in alkaline, salt-laden soils. It's also grown as a firebreak, as the high salt content in the canopy makes it difficult to burn. Unlike the other introduced members of the genus, which are costly, invasive weeds growing in watercourses and washes throughout the southwestern United States, the athel tree mostly makes sterile seeds and rarely escapes cultivation.

These days are the hard times.
Some residents depend on the
forest for their survival.
—Marcos Malubay

Branches with
many scalelike,
minute leaves

Branches can be
mistaken for pine
needles

Tilia spp.

TILL-ee-ah

Tilia - L., linden trees

Malvaceae

North Temperates

Simple, Alternate

Deciduous, 25–50 ft.

T. cordata

Tilia is a genus of about 30 species of trees native to temperate areas of the Northern Hemisphere. They tend to grow best in fertile soils in areas with cold winters, so in California, they are only commonly found in Central Valley towns and cities. The genus is easily recognized by its lopsided, heart-shaped, toothed leaves and by its fragrant, yellowish white, five-petaled flowers. The flowers, and later the fuzzy, pea-shaped fruits, are attached to a conspicuous, tongue-shaped bract (a modified leaf) that serves as a wind-dispersal agent, detaching with the fruit and fluttering, like a helicopter, away from the parent tree.

Silver Linden
(*T. tomentosa*)
leaf and flowers

Largeleaf linden
(*T. platyphyllos*)
leaf, bracts,
developing
flowers,
and fruits

Bracts

Key to California's Commonly Cultivated *Tilia*

1. Leaf underside white, densely hairy—Silver Linden (*T. tomentosa*)
1' Leaf underside blue-green (glaucous) or green, hairless, thinly hairy, or hairy only along veins and in vein axils
    2. Leaves blue-green (glaucous) beneath—Littleleaf Linden (*T. cordata*)
    2' Leaves green beneath
        3. Leaves sparsely hairy beneath, particularly along veins—
            Largeleaf Linden (*T. platyphyllos*)
        3' Leaves hairless beneath except for tufts of hair in vein axils
            4. Leaves without tufts of hair in vein axils at leaf base; some
                stamens modified into petal-like staminodes that lack anthers—
                American Linden (*T. americana*)
            4' Leaves with tufts of hair in vein axils at leaf base; all stamens
                similar and bearing anthers—Crimean Linden (*T. × euchlora*)

Consider the life of trees. Aside from the axe, what trees acquire from man is inconsiderable. What man may acquire from trees is immeasurable.
—*Cedric Wright*

Tipuana tipu Tipu Tree

TEE-poo-ah-nah TEE-poo
Tipuana tipu - Tipuani Valley in Bolivia
Fabaceae

Brazil, Bolivia, Argentina
Pinnate, Opposite and Alternate
Briefly Deciduous, 30–40 ft.

The tipu tree exhibits two traits that make it unique and easily recognized among other commonly grown trees in the pea family. Its large leaves (pinnately compound, with 5 to 10 pairs of leaflets and a single terminal leaflet) are often borne in pairs (oppositely) on the stem, whereas the leaves of most other legumes arise individually (alternately). Also, its distinctive fruit is a highly modified bean pod with a flattened wing; it spins like a helicopter when falling from the tree. The rapidly growing tipu tree is planted throughout warmer parts of California. It is praised for its beauty, drought tolerance, and durability in varying soil conditions, but it is criticized for having invasive roots and brittle wood. The fallen remnants of its lovely, apricot orange to lemon yellow flowers carpet sidewalks during early summer.

If you reveal your secrets to the wind you should not blame the wind for revealing them to the trees. —Kahlil Gibran

Fruit

Flowers

Triadica sebifera

Chinese Tallow Tree

try-ADD-dih-kuh seh-BIFF-er-ah
Trias - Gr., in threes
sebifera - L., tallow-bearing
Euphorbiaceae
Synonym: *Sapium sebiferum*

China, Northern Vietnam
Simple, Alternate
Deciduous, 30–40 ft.

Many trees require cold temperatures to display good fall foliage color. However, even in the moderate chill afforded most trees in California, the Chinese tallow tree still displays amazing golden, orange, crimson, and plum purple fall color. On occasion, all of California's fall colors can be found in a single leaf, swirled and coursing together in deliquescent lines. This species, previously known as *Sapium sebiferum*, is distinguished from trees in the *Sapium* genus by the two swollen glands at the attachment of the leaf blade to the leaf stalk. The Chinese tallow tree, which is an aggressive weed in forests, fields, and coastal prairies of the southeastern U.S., can be recognized by its scaly gray trunk, poplar-like leaves, milky poisonous sap, and spikes of small yellow flowers formed on branch ends. The female flowers develop into three-chambered capsules that open to reveal seeds with a white, waxy seed coat. The seed coat, which is rich in saturated fat, is processed in Asia for candle and soap making.

Leaf base showing two glands

Variation in fall color

Fruit capsules with waxy, white seeds

I think people tend to forget that trees are living creatures. They're sort of like dogs. Huge, quiet, motionless dogs, with bark instead of fur. —Jack Handy

Trees with Great Fall Color in California

Japanese Maple *(Acer palmatum)*: Red
Maidenhair Tree *(Ginkgo biloba)*: Golden
Crape Myrtle *(Lagerstroemia hybrids)*: Many colors
Sweetgum *(Liquidambar styraciflua)*: Many colors
Tulip Tree *(Liriodendron tulipifera)*: Golden orange
White Mulberry *(Morus alba)*: Yellow
Chinese Pistache *(Pistacia chinensis)*: Red and orange
Callery Pear *(Pyrus calleryana)*: Red
Scarlet Oak *(Quercus coccinea)*: Red
Chinese Tallow Tree *(Triadica sebifera)*: Many colors
Sawleaf Zelkova *(Zelkova serrata)*: Red and orange

Tristaniopsis laurina   Water Gum

tris-tan-ee-OP-sis lor-EE-nah

Tristania - Jules Tristan (1776–1861); *opsis* - Gr., likeness

laurina - L., resembling a laurel tree

Myrtaceae

Synonym: *Tristania laurina*

Eastern Australia
Simple, Alternate
Evergreen, 30+ ft.

Fruits on the stem

Water gum has a wide range in its eastern Australian native habitat, growing along coastal watercourses in Queensland, New South Wales, and Victoria, where it varies from a small shrub to a tree exceeding 50 feet in height. In cultivation, the water gum remains a tidy tree with a dense, dark green crown of glossy leaves and smooth, scaly, shedding bark reminiscent of *Eucalyptus*. Holding one of the leaves, which are darker above and pale, yellowish green beneath, to the light will reveal many transparent oil dots—a characteristic of many plants in the myrtle family. In late spring and early summer, water gum produces a profusion of small, five-petaled, bright yellow flowers. This species is popular on the streets of milder California cities, especially San Francisco, and with good reason: it withstands wind, poor soils, disease, and drought, drops negligible litter, and requires little pruning or care after becoming established. Its modest stature and slow growth also make it a suitable choice for planting under power lines.

Bark

Flower and
dotted leaves

*Trees are Earth's endless effort
to speak to the listening heaven.*
—Rabindranath Tagore

Ulmus parvifolia

Chinese Elm

ULL-mus par-vi-FOE-lee-ah
Ulmus - L., elm
parvifolia - L., small-leaved
Ulmaceae

China, Japan, and Korea
Simple, Alternate
Deciduous/Evergreen, 30–60 ft.

Fruits (samaras)

The most common elm in California's urban and suburban environments is the Chinese elm. This species is admired for its fast growth, resistance to Dutch elm disease, and graceful, wide-spreading, weeping form. It can be distinguished from other elms by its scaly, mottled, gray trunk, with outer bark that flakes off in patches, exposing orange and olive green interior bark. Chinese elm leaves are mostly evergreen in warmer coastal areas and partially or fully deciduous in colder, interior cities.

And this, our life, exempt from public haunt, finds tongues in trees and good in everything. —William Shakespeare

Singly serrate leaf

Bark

Ulmus spp.   Other Elms

Ulmus - L., elm
Ulmaceae

North Temperates
Simple, Alternate
Deciduous, 30–100 ft.

T he 30 species of elms are mostly distributed throughout the north temperates. Although no elms are native to the western U.S., several species are commonly cultivated in California for their fine foliage and dignified forms. Due to their many hybrids and cultivated varieties, elms can be difficult to identify. The flowers and fruits are small and inconspicuous, and the leaves are highly variable. However, the two most salient characteristics for identifying many members of the genus are doubly serrate leaf margins and more or less lopsided leaf-blade bases. Tiny flowers without petals usually emerge before the new leaves in spring and develop into flattened, nutlike fruits (called "samaras") surrounded by a round, papery wing.

U. minor

U. americana
trunk

U. americana
fall leaf color

Key to Commonly Cultivated Elms

1. Leaf margins mostly singly serrate, nearly symmetrical at the base
    2. Leaves evergreen or partly so, mostly <1 in. wide; bark smooth and flaking; fruit appearing in fall—Chinese Elm *(U. parvifolia)*
    2′ Leaves deciduous, mostly 1 in. wide or more; bark rough; fruit appearing in spring—Siberian Elm *(U. pumila)*
1′ Leaf margins at least partially doubly serrate, asymmetrical at the base
    3. Leaf upper surface smooth to the touch—Dutch Elm *(U. × hollandica)*
    3′ Leaf upper surface rough to the touch
        4. Leaf upper surface only slightly rough to the touch, tufts of hair only occasionally in lower surface vein axils; fruit margins hairy—American Elm *(U. americana)*
        4′ Leaf upper surface very rough to the touch, tufts of hair always in lower surface vein axils; fruit margins hairless
            5. Leaf base strongly lopsided, one side sometimes covering the petiole, seed in the center of the samara, leaves with 16 to 20 pairs of veins—Scotch Elm *(U. glabra)*
            5′ Leaf base lopsided but not strongly so, petiole not covered, seed toward the tip of the samara, leaves with 8 to 18 pairs of veins—English Elm *(U. minor*, synonym = *U. procera)*

Doubly serrate U. americana leaf with asymmetrical base

Umbellularia californica California Bay Laurel

um-bell-yew-LARE-ee-ah kal-ih-FOR-ni-kah

Umbellularia - L., little umbrella

californica - L., California

Lauraceae

☿

Southwestern Oregon to Baja California

Simple, Alternate

Evergreen, 40–70 ft.

The native California bay laurel is usually associated with sycamores and oaks in shaded watercourses and oak woodlands. It also grows in neighborhood parks and open spaces at the interface between the state's wild and urban areas, and occasionally as a garden specimen. It bears six-part, yellowish-white flowers in umbrella-shaped clusters from winter to early spring, followed by olive-sized fruits that ripen to a wrinkled purple. The fine-grained, honey-colored wood, sometimes called Oregon myrtle, is used in furniture, bowl turning, and carving. The dark green, pungent-smelling leaves can replace bay leaves (*Laurus nobilis*) in stew and soup flavoring. This species is a primary host of sudden oak death (*Phytophthora ramorum*) and can transmit the disease to oaks and tanoak without succumbing to it.

Fruits

Flowers

Trees can reduce the heat of a summer's day, quiet a highway's noise, feed the hungry, provide shelter from the wind and warmth in the winter. —George H. W. Bush

Zelkova serrata   Sawleaf Zelkova

zell-KOE-vah ser-RAW-tah
Zelkova - local name for Caucasian Elm
  (Z. carpinifolia)
serrata - L., sawlike teeth
Ulmaceae

Japan, Korea, Taiwan, China
Simple, Alternate
Deciduous, 80 ft.

Leaves
and fruits

The graceful, funnel-shaped sawleaf zelkova is planted extensively in the eastern and midwestern U.S. for its brilliant orange, red, and yellow fall color, often replacing American elms lost to Dutch elm disease. This species is rarely grown in coastal Southern or Central California, yet it is one of the most common shade trees on the avenues of Central Valley towns. It can be recognized by its coarsely toothed, elmlike leaves with upper surfaces that feel like sandpaper, its mottled gray trunk with light brown and orange patches, and its small, brown fruit, nestled in the axils of mature leaves. The moisture resistant, fine-grained wood is prized in Japan, where it is used especially in temple construction.

Fruits

A tree says: a kernel is hidden in me, a spark, a thought, I am life from eternal life. The attempt and the risk that the eternal mother took with me is unique, unique the form and veins of my skin, unique the smallest play of leaves in my branches and the smallest scar on my bark. I was made to form and reveal the eternal in my smallest special detail.
—Herman Hesse

Bark

It is not so much for its beauty that the forest makes a claim upon men's hearts, as for that subtle something, that quality of air, that emanation from old trees, that so wonderfully changes and renews a weary spirit.
—Robert Louis Stevenson

Palms

Palms

Although, unlike broad-leaved trees and conifers, they do not grow annual rings of wood, palms are trees nonetheless, often with large, single trunks and sizable canopies. The 2,600 or so species of palms are among the wonders of the natural world. Most are frost-tender, from subtropical and tropical parts of the world, with only a few species occurring in upper latitudes, such as the Mediterranean fan palm (*Chamaerops humilis*), Chilean wine palm (*Jubaea chilensis*), and California's only native palm, the California fan palm (*Washingtonia filifera*). Very few palms make branches; they mostly have an undivided, single trunk, which does not thicken with age, and an all-important apical bud from which all the leaves arise at the growing tip. If this bud is cut out or frozen, the entire tree will die. Some palms can grow to be extremely tall, like the Andean wax palm (*Ceroxylon* spp.), which reaches nearly two hundred feet. The vining rattan palms (*Calamus* spp.) have the longest stems in the world, some nearly six hundred feet long. Palms have a strong, grasslike, fibrous root system lacking a taproot, and roots often arise beneath the bark on the base of the trunk (see the widening bases of old Canary Island date palms [*Phoenix canariensis*]). Humans make use of palms in a number of ways. The trees' apical buds are harvested as palm hearts, their sap is drained for fermentation (Chilean wine palm), and their leaves, stems, and trunks are used in roof thatching, furniture making, weaving, and construction. Foods come from the energy-rich fruits of many species: dates (*Phoenix dactylifera*), acai (pronounced awe-saw-EE) (*Euterpe oleracea*), palm oil (*Elaeis guineensis*), and coconut (*Cocos nucifera*).

Pigmy Date Palm
(*Phoenix roebelenii*) fruits

Mediterranean Fan Palm
(*Chamaerops humilis*)

Canary Island Date Palms
(*Phoenix canariensis*)

Chilean Wine Palm
(*Jubaea chilensis*)

Curly Palm
(*Howea belmoreana*)

Key to California's Commonly Cultivated Palms

California's most commonly grown palm trees (bolded) are treated in detail on the following pages.

1. Leaves bipinnately divided—Fishtail Palm (*Caryota* spp.)
1' Leaves pinnately divided or palmately divided
  2. Leaves pinnately divided (feather palms)
    3. Lower leaflets becoming sharp spines
      4. Trunk diameter >1.5 ft.
        5. Leaves pale green or bluish green, erect and ascending—Date Palm (*Phoenix dactylifera*)
        5' Leaves dark green, strongly arched—**Canary Island Date Palm (*Phoenix canariensis*)**
      4' Trunk diameter <1.5 ft.

Fishtale Palm
(*Caryota gigas*)

        6. Leaves <4 ft. long—Pigmy Date Palm (*Phoenix roebelenii*)
        6' Leaves >5 ft. long—Senegal Date Palm (*Phoenix reclinata*)
    3' Lower leaflets not spines
      7. Trunk diameter >3 ft.—Chilean Wine Palm (*Jubaea chilensis*)
      7' Trunk diameter <2 ft.
        8. Flower clusters arising from below the green crown shaft of the trunk, >2 ft. below the leaves—**King Palm (*Archontophoenix cunninghamiana*)**
        8' Flower clusters arising from among the leaves
          9. Petiole margins with spiny black teeth, leaves gray-green—Pindo Palm (*Butia capitata*)
          9' Petiole margins unarmed, leaves green

Edible Pindo Palm
(*Butia capitata*) fruits

Even trees do not die without a groan. —Henry David Thoreau

          10. Trunk >15 in. diameter; leaflets feather-duster like, arising from central leaf axis in multiple planes, bent and drooping at the tip—**Queen Palm (*Syagrus romanzoffiana*)**
          10' Trunk <15 in. diameter; leaflets emanating from central leaf axis in a single plane, not bent at the tip—Kentia Palm (*Howea* spp.)
  2' Leaves palmately divided (fan palms)
    11. Petiole margins conspicuously armed with spiny teeth, at least at the base
      12. Low, clumping palm with multiple stems from the ground—Mediterranean Fan Palm (*Chamaerops humilis*)
      12' Treelike palm with single trunk
        13. Old leaves not forming a skirt around the trunk, leaflet tips conspicuously bent and drooping, petals not flat, with grooves on inner surface—Australian Fan Palm (*Livistona* spp.)

Guadalupe Palm
(*Brahea edulis*) fruits

        13' Old leaves forming a skirt around the trunk (unless pruned off), leaflet tips mostly erect, petals thin and flat, without grooves on inner surface
          14. Trunk stout (usually >2 ft. in diameter), many hairlike fibers on the leaf tips—**California Fan Palm (*Washingtonia filifera*)**
          14' Trunk very tall and slender (usually <2 ft. in diameter), few hair-like fibers on the leaf tips—**Mexican Fan Palm (*Washingtonia robusta*)**
    11' Petiole margins unarmed or occasionally with small, sharp teeth
      15. Trunk covered with coarse brown fibers and remnants of old leaf bases—**Windmill Palm (*Trachycarpus fortunei*)**
      15' Trunk smooth and gray, without coarse fibers
        16. Leaves silvery gray to blue-green—Blue Fan Palm (*Brahea armata*)
        16' Leaves green—Guadalupe Palm (*Brahea edulis*)

Archontophoenix cunninghamiana   King Palm

ar-kon-toe-FEE-niks kuh-ning-ham-ee-AYE-nah

Archontos - Gr., chieftain; *phoenix* - Gr., date palm

cunninghamiana - James Cunningham (d. 1709)

Arecaceae

Eastern Australia
Pinnate, Alternate
Evergreen, 40 ft.

Flowers

The king palm only survives in areas that are frost-free; mature trees can die at around 25°F. It is popular as a container-grown palm in shopping malls in Northern California. The king palm has a clean and stately appearance with leaves like those of coconut palms gracefully arching from atop a smooth, gray trunk ringed with old leaf scars. Heavily branched clusters of lavender flowers hang from the trunk below the leaves and are followed by a splendid display of jewel-like, red fruits. This species is rare and protected in its native Australian habitat, yet it is considered an invasive exotic in Brazil. The king palm is often mistakenly sold under the scientific name *Seaforthia elegans*, which is an altogether different palm, also from Australia. The closely related Alexandra palm (*A. alexandrae*), which has a comparatively larger trunk base that tapers upward to the leaves, is grown much more commonly in Florida than·in California.

Fruits

A giant tree, protected only by its trunk, is capable of being destroyed in half an hour. Such vulnerability is disturbing to those who love trees. —Francis Hallé

Phoenix canariensis   Canary Island Date Palm

FEE-nicks kah-nair-ee-EN-sis
Phoenix - Gr., date palms
canariensis - L., of the Canary Islands
Arecaceae

Canary Islands
Pinnate, Alternate
Evergreen, 75 ft.

The native home of this date palm, the Canary Islands, is a small archipelago off the western coast of Morocco. This spectacular tree was originally brought to California from Southern Europe by the mission fathers, and relic individuals, planted long ago, still persist near missions and early California homesteads. The Canary Island date palm can be identified by its thick trunk covered in diamond-shaped leaf scars, splays of inedible, orange fruits, and resplendent crown of over one hundred feather-shaped fronds. This crown can be home to entire ecosystems of rodents, birds, insects, and plants—fig trees (*Ficus carica*) planted by birds often grow thirty feet off the ground in these palms. As anyone who has ever tried to trim the twenty-foot-long leaves from one of these trees knows, the lower leaflets grow into stout spines. A close relative, the date palm (*P. dactylifera*), the source of commercial dates, can be distinguished by its more slender trunk and more upright, bluish gray leaves.

Fruits

Leaf scars on the trunk

California's "Old-Timey" Trees

Trees planted in California long ago and now regularly found near old home sites and missions

Tree of Heaven *(Ailanthus altissima)*
Bunya Bunya Tree *(Araucaria bidwillii)*
Blue Gum *(Eucalyptus globulus)*
Monterey Cypress *(Hesperocyparis macrocarpa)*
Southern Magnolia *(Magnolia grandiflora)*
Olive *(Olea europaea)*
Canary Island Date Palm *(Phoenix canariensis)*
Monterey Pine *(Pinus radiata)*
Black Locust *(Robinia pseudoacacia)*
Pepper Tree *(Schinus molle)*

Planting a tree is taking a stand against desolation. It is believing in life and beauty. It is investing in the future.
—*Jeffrey Meyer*

Syagrus romanzoffiana

Queen Palm

sye-AG-rus roe-man-zoff-ee-AN-ah

Syagrus - L., name used by Pliny for a palm

romanzoffiana - Count Nicholas
de Romanzoff (1754–1826)

Arecaceae

Brazil to Northeast Argentina

Pinnate, Alternate

Evergreen, 50 ft.

A bold palm with a tropical air, the queen palm is widely grown throughout the subtropical world, especially in Australia and Southern California, where it is the most prevalent palm on many city streets. It can be readily identified, as it is the only palm in California with leaves shaped like feather dusters, with leaflets emanating from a central axis in all directions. Other palms have palmately divided, fan-shaped leaves (i.e., *Washingtonia* spp.) or pinnately divided, feather-shaped leaves with leaflets only in one plane (i.e., *Phoenix canariensis*). The smooth trunk of the fast-growing queen palm looks like a telephone pole ringed at one-foot intervals with light gray leaf scars. Thousands of flowers, which allure bees, emerge in hanging clusters from the leaf bases and develop into orange, edible fruits.

Flower
and fruit
clusters

Fruits

A tree is an incomprehensible mystery. —Jim Woodring

Trachycarpus fortunei

Chinese Windmill Palm

tray-kee-KAR-puss for-TOON-ee-eye
Trachys - Gr., rough; *karpos* - Gr., fruit
fortunei - Robert Fortune (1812–1880)
Arecaceae

Northern Myanmar, China, Japan
Palmate, Alternate
Evergreen, 30 ft.

Many trees naturalize (reproduce on their own) after being carried by humans outside their native range. Introductions are not recorded, time passes, trees reproduce, and it becomes difficult to determine if a species is native or naturalized in a given area. This is especially true of economically and culturally important plants with a long history of cultivation, such as olives, figs, and ginkgos. The Chinese windmill palm is thought to be native to mountainous areas of Northern Myanmar, central and southern China, and possibly Japan, but it has been cultivated for its many uses for so long that its true native range is unclear. A hardy palm, it is extensively planted along streets, in parks, and in gardens throughout California and survives as a street tree in adverse conditions. Its downward-tapering trunk is covered with coarse brown fibers and remnants of old leaf bases. The nearly round, fanlike leaves are divided almost to the middle into narrow leaflets and attached to the trunk by slender, finely-toothed stalks. In spring, clusters of edible, pale yellow flowers develop into inedible bruise-colored fruits shaped like plump kidneys.

Fruits

Trees—they stand erect, they are hard, they are virile. —Francis Hallé

Washingtonia filifera

wah-shing-TOE-nee-uh fih-LIH-fer-uh
Washingtonia - George Washington
(1732–1799)
filifera - L., thread-bearing
Arecaceae

Southeastern California,
Sonoran Desert oases
Palmate, Alternate
Evergreen, 60 ft.

There are only two species in the genus *Washingtonia* and both can be found as street and park trees throughout most of California. The two species can be difficult to distinguish, and hybrids between the two exist in cultivation. In general, *W. filifera* is shorter, with a stouter trunk, more hairlike fibers on the leaf tips, and a less compact crown with longer leaf stalks (petioles) than *W. robusta*. The large, palmate leaves of both species are armed on the petiole with prominent, hooked spines. These protective teeth become reduced or absent when the trees reach thirty to forty feet, a height at which they would have been out of reach of the large, extinct herbivores of the distant past. California fan palm is the only palm native to California, where it grows near desert streams, springs, and arroyos. Depending on the level of pruning, the trunk is either smooth or covered with a cross-hatched pattern of persistent leaf bases. Untrimmed trees can have a large skirt of dead leaves, occasionally covering the entire trunk (hence the other common name, "petticoat palm").

Spines on the leaf petiole

Except during the nine months before he draws his first breath, no man manages his affairs as well as a tree does. —George Bernard Shaw

Washingtonia robusta   Mexican Fan Palm

wah-shing-TOE-nee-uh roe-BUS-tuh

Washingtonia - George Washington
  (1732–1799)

robusta - L., stout

Arecaceae

Baja California and Sonora, Mexico

Palmate, Alternate

Evergreen, 100 ft.

Mexican fan palm is the most widely grown palm tree in coastal California (and probably the entire U.S.), where nary an ocean view lacks their characteristic silhouette accenting the horizon. Old specimens have astonishing proportions, with a graceful trunk up to one hundred feet tall while only about two feet wide (a fifty-to-one height-to-diameter ratio). Few other plants are so tall and thin. The height-to-diameter ratio of the world's tallest non-palm trees, such as the coast redwood (*Sequoia sempervirens*), rarely exceeds fifteen to one.

Accordion-like leaf
folds of the large,
fan-shaped leaves

Fruits

As the poet said, "Only God can make a tree"—probably because it's so hard to figure out how to get the bark on.
—Woody Allen

Appendix A: *Changing Plant Names*

There are two different kinds of plant names: common names and scientific names. Common names used in this book are those in the *Sunset Western Garden Book*. However, common names can be misleading: some plants have more than one common name and some common names apply to more than one kind of plant (see the discussion of cedars on page 32).

Species names consist of two words, written in Latin: the first is the genus in which the plant is classified, and the second, the specific epithet, is usually a Latin adjective (e.g., *biloba,* meaning two-lobed) or a possessive noun (e.g., *bidwillii,* honoring John Carne Bidwill). Scientific names are assigned to plants based on their relatedness. For instance, all species in the same genus are more closely related to each other than they are to species in other genera.

Scientific names, however, are not static and can change as new information about relationships comes to light. New studies can lead to a better understanding of the relationships among different species. This new understanding and a desire for plant names to represent true evolutionary relationships have led to numerous scientific name changes, even in some of California's most commonly grown trees. Scientific names also follow the rule of priority of publication, in which the first correctly published name for a plant is the one used. Occasionally an earlier published name is discovered that leads to a name change (e.g., *Platanus* × *hispanica* published in 1770 and *Platanus* × *acerifolia* published later, in 1805).

Common Name	Scientific Name Used in This Book	Synonym
African Fern Pine	*Afrocarpus falcatus*	*Podocarpus gracilior*
Floss Silk Tree	*Ceiba speciosa*	*Chorisia speciosa*
Lemon Scented Gum	*Corymbia citriodora*	*Eucalyptus citriodora*
Red Flowering Gum	*Corymbia ficifolia*	*Eucalyptus ficifolia*
Trumpet Trees	*Handroanthus* spp.	*Tabebuia* spp.
Monterey Cypress	*Hesperocyparis macrocarpa*	*Cupressus macrocarpa*
Brisbane Box	*Lophostemon confertus*	*Tristania conferta*
London Plane Tree	*Platanus* × *hispanica*	*Platanus* × *acerifolia*
African Sumac	*Searsia lancea*	*Rhus lancea*
Japanese Pagoda Tree	*Styphnolobium japonicum*	*Sophora japonica*
Chinese Tallow Tree	*Triadica sebifera*	*Sapium sebiferum*
Water Gum	*Tristaniopsis laurina*	*Tristania laurina*

Appendix B: *Locations of Photographed Trees*

Acacia baileyana, 17th & Shrader, San Francisco

Acacia melanoxylon, Franklin & California, San Francisco

Acacia stenophylla, Skylark Lane, San Luis Obispo

Acer palmatum, Highland Dr., Santa Cruz

Aesculus hippocastanum, Divisidero & Vallejo, San Francisco

Afrocarpus falcatus, Mission Park, Ventura

Ailanthus altissima, Garden & Pismo, Backlot, San Luis Obispo

Albizia julibrissin, Dora & Hope, Ukiah

Alnus rhombifolia, Dahlia Lane, San Luis Obispo

Araucaria bidwillii, Cal Poly, San Luis Obispo

Arbutus 'Marina', Pacific & Baker, San Francisco

Archontophoenix cunninghamiana, Plaza Rubio, Santa Barbara

Bauhinia variegata, Olive & De La Guerra, Santa Barbara

Betula pendula, Jeffrey & Highland, San Luis Obispo

Brachychiton acerifolius, Marsh & Chorro, San Luis Obispo

Brachychiton rupestris, Los Angeles County Arboretum

Callistemon viminalis, Garden & Buchon, San Luis Obispo

Calocedrus decurrens, University of California, Berkeley

Caryota gigas, Santa Barbara & Yanonoli, Santa Barbara

Cassia leptophylla, 41st & Portola, Capitola

Casuarina spp., Lincoln & University, Palo Alto

Catalpa spp., Addison & High, Palo Alto

Cedrus deodara, Healdsburg

Ceratonia siliqua, Johnson & Palm, San Luis Obispo

Cercis canadensis, University of California, San Diego

Chamaerops humilis, San Luis Drive & El Centro, San Luis Obispo

Chionanthus retusus, Cal Poly Campus, San Luis Obispo

Cinnamomum camphora, Harvest & Pajaro, Salinas

×Chitalpa tashkentensis, Walnut Creek

Corymbia ficifolia, Osos & Leff, San Luis Obispo

Crataegus phaenopyrum, Capitola & Maciel, Santa Cruz

Cupaniopsis anacardioides, Manhattan Ave., Manhattan Beach

Cupressus sempervirens, California State University, Chico

Eriobotrya deflexa, Mitchell Park, San Luis Obispo

Erythrina caffra, San Vincent, Santa Monica

Erythrina × sykesii, Leaning Pine Arboretum, San Luis Obispo

Eucalyptus cladocalyx, Santa Barbara Mission, Santa Barbara

Eucalyptus globulus, University of California, Berkeley

Eucalyptus nicholii, Nipomo & Buchon, San Luis Obispo

Ficus macrophylla, South Coast Botanical Garden, Palos Verdes; Balboa Park, San Diego; Carroll Ave. & Douglas St., Angelino Heights; Beverly Dr. & Park Way, Beverly Hills

Ficus microcarpa, Santa Monica Blvd. & Crest, Beverly Hills; Pacific Heights, San Francisco

Ficus rubiginosa, UC Irvine campus

Fraxinus angustifolia 'Raywood', El Camino Real & Traffic Way, Atascadero

Fraxinus uhdei, Alpine Dr., Beverly Hills; Greta Pl., San Luis Obispo

Fraxinus velutina 'Modesto', Sydney Dr., San Luis Obispo

Geijera parviflora, Laguna & De La Guerra, Santa Barbara

Ginkgo biloba, Walnut & Chestnut, Santa Cruz

Gleditsia triacanthos, Main & Palm, Ventura

Grevillea robusta, Santa Barbara & Leff, San Luis Obispo

Handroanthus chrysotrichus, Alice Keck Park, Santa Barbara

Hespercyparus macrocarpa, Golden Gate Park, San Francisco

Heteromeles arbutifolia, Cal Poly Campus, San Luis Obispo

Howeia belmoreana, Plaza Rubio, Santa Barbara

Hymenosporum flavum, Los Verdes Dr., San Luis Obispo

Jacaranda mimosifolia, Pacific & Johnson, San Luis Obispo

Jubaea chilensis, Spencer Adams Park, Santa Barbara

Juglans regia, Chico

Juniperus chinensis, Laguna & Plaza Rubio, Santa Barbara

Koelreuteria bipinnata, University of California, San Diego

Lagerstroemia 'Tuskegee', 16th & H, Sacramento

Laurus nobilis, Garden & Micheltorena, Santa Barbara

Ligustrum lucidum, State & Scott, Ukiah

Liquidambar styraciflua, Ocean View Ave., Santa Cruz

Liriodendron tulipifera, Oak & Seminary, Ukiah

Lophostemon confertus, Marsh & Garden, San Luis Obispo

Lyonothamnus floribundus subsp. *aspleniifolius,* Buchon St., San Luis Obispo

Magnolia grandiflora, Forrest & Oleander, Bakersfield

Malus 'Liset', San Francisco Botanical Garden, San Francisco

Maytenus boaria, Maple & Center, Santa Cruz; Baker and North Point, San Francisco

Melaleuca linariifolia, Palm & Chorro, San Luis Obispo

Melaleuca quinquenervia, Milpas & Calle Puerto Vallarta, Santa Barbara

Melia azedarach, 20th & D, Bakersfield

Metasequoia glyptostroboides, Lotusland, Montecito; Hamilton Avenue, Palo Alto; Capitol Gardens, Sacramento

Metrosideros excelsa, Cole Valley, San Francisco

Michelia doltsopa, Johnson & Pepper, San Luis Obispo

Morus alba, Morro & Buchon, San Luis Obispo

Myoporum laetum, Azure St., Morro Bay

Olea europaea, Gilroy, San Luis Obispo, San Francisco, Stanford University

Parkinsonia × 'Desert Museum', Huntington Botanical Gardens, San Marino

Phoenix canariensis, Hillcrest & Carmelita, Beverly Hills; Alameda Park, Santa Barbara

Pinus canariensis, Lexington Rd., Beverly Hills

Pinus pinea, Pacific & Morro, San Luis Obispo

Pinus radiata, High St. & Emmett St., Santa Cruz

Pistacia chinensis, Saratoga Ave., Monte Sereno

Pittosporum undulatum, Cal Poly, San Luis Obispo

Platanus × *hispanica,* Water & River, Santa Cruz; Getty Museum Café, Los Angeles

Podocarpus macrophyllus, Micheltorena & State, Santa Barbara

Populus nigra 'Italica', University of California, Davis; Auburn & Benicia, Milpitas

Prunus campanulata, Cal Poly, San Luis Obispo

Prunus cerasifera, Chestnut & California, Salinas; McGee & Berkeley Way, Berkeley

Pyrus calleryana, Orchard & W. Trimble, San Jose; Westchester, Bakersfield

Pyrus kawakamii, Balboa Park, San Diego

Quercus agrifolia, Corrida & Woodbridge, San Luis Obispo

Quercus ilex, Echo St., Fresno

Quercus lobata, Buchon & Garden, San Luis Obispo

Quercus rubra, Cedar St. & Walnut Ave., Santa Cruz

Quercus suber, University of California, Davis

Robinia × *ambigua,* Palo Alto

Schinus molle, Stanford University, Palo Alto

Searsia lancea, Bay & California, Santa Cruz

Sequoia sempervirens, Capitol Gardens, Sacramento

Spathodea campanulata, Encinitas

Stenocarpus sinuatus, Prado & Balboa Dr., Balboa Park, San Diego

Styphnolobium japonicum, Big Basin Way, Saratoga

Syagrus romanzoffiana, San Diego State University, San Diego

Syzygium australe, Olive & Sola, Santa Barbara

Tamarix aphylla, Tamarisk Grove Campground, Anza Borrego Desert State Park

Tilia cordata, Capitol & 10th, Sacramento

Trachycarpus fortunei, Junipero Plaza St., Santa Barbara

Triadica sebifera, Via Mimosa, The Villages, San Jose

Tristaniopsis laurina, Main & Ford, Watsonville

Ulmus minor, California State University, Chico

Ulmus parvifolia, Santa Monica Blvd. & Crest, Beverly Hills

Washingtonia filifera, Stanford campus; N & 15th, Sacramento

Washingtonia robusta, Ocean Front & 18th, Venice Beach; Channel Drive, Montecito

Zelkova serrata, Sacramento

Glossary

Acorn The fruit of an oak tree. A thin-walled nut that sits in a cup of modified leaves.

Alternate leaf arrangement One leaf is attached per node on the stem.

Angiosperm A plant that makes flowers and fruit.

Anther The enlarged, pollen-forming portion of a stamen, borne at the tip of the filament.

Arboretum An institution that collects, grows, studies, and exhibits trees.

Asymmetric Having parts that are not divisible into identical or mirror-image halves.

Awl-like leaf Narrow, triangular, sharp-pointed, often thick and usually less than ½ inch long.

Axil The upper angle between stem and leaf or leaf stalk (petiole).

Bark Outer layers of tissue covering the wood.

Basal Found at or near the base (proximal).

Berry A fleshy, many-seeded fruit.

Bipinnate Compound leaf form in which the primary leaflets are again divided into secondary leaflets.

Bisexual Refers to a flower containing both male and female organs.

Blade The expanded photosynthetic portion of a leaf.

Bract A small, modified leaflike structure associated with reproductive parts, such as inflorescences or flowers.

Branchlets Small, leaf-bearing branches or twigs, usually on the most recent growth.

Bud An immature shoot that has the potential to develop into a branch, flower, or inflorescence, found in the axil of a leaf and/or the terminal tip of a shoot, often covered with protective scale leaves.

Calyx The collective term for all the sepals of a flower: the outermost whorl of flower parts, typically green.

Capsule A dry, often many-seeded fruit that splits apart at maturity (dehiscent).

Carpel The basic female structure of flowering plants, in which the ovules are borne. A number of carpels can be fused to form a pistil.

Catkin A spikelike, often pendent cluster of flowers that are all the same sex.

Character A feature or attribute of a plant, or a funny person.

Columnar In the form of a column.

Compound leaf A leaf whose blade is divided into distinct leaflets.

Cone The reproductive structure of conifers (and other nonflowering plants) with the scales overlapping (a strobilius). Also refers to any inflorescence or fruit with overlapping scales.

Conifer A cone-bearing tree, such as a pine or juniper.

Corolla The collective term for all the petals of a flower: the whorl of flower parts just inside the calyx.

Crown The top of a tree, including the upper trunk and branches.

Cultivar A cultivated variety of a plant.

Deciduous Said of plant structures that fall off naturally at the end of a growing period, e.g., leaves that fall seasonally or plants that are seasonally leafless.

Dehiscent Splitting open at maturity.

Dioecious Refers to species in which male and female reproductive structures are produced on separate individuals.

Disturbed area A weedy waste area that has been ecologically disrupted or destroyed by human activities or natural processes such as fire.

Doubly serrate Said of a leaf margin whose serrations bear smaller teeth on their margins (small teeth on larger teeth).

Endangered species A species whose survival is in immediate jeopardy.

Endemic species A species that occurs only in a defined area.

Entire Said of a leaf whose margins are continuous and smooth, toothless.

Evergreen A plant that has leaves that remain attached, i.e., do not fall off all at once.

Family A taxonomic rank. From most to least inclusive, the ranks are domain, kingdom, phylum, class, order, family, genus, and species. Family names end in "-aceae."

Fascicle A bunch or tuft of leaves all arising from the same place.

Filament The threadlike, stalk portion of a stamen, atop which sits the anther.

Flower The reproductive part in angiosperms, consisting of some combination of the sepals, petals, stamens, and carpels.

Follicle A dry fruit that splits open on only one side.

Fruit The ripened ovary that protects the seeds and aids in their dispersal.

Fused Refers to structures that are united, at least at the base, sometimes throughout.

Genus (plural: genera) A group of related species; the taxonomic category that ranks above "species" and below "family."

Glabrous Without hairs.

Gland A small, often spheric body on or embedded in the epidermis or at the tip of a hair.

Glandular Having glands.

Glaucous Covered with a whitish or bluish film that is waxy or powdery.

Grafting A horticultural technique by which a bud of one plant (parent) is bonded to a rooted plant (root stock), thereby creating a clone of the parent.

Gymnosperm A plant that makes woody or fleshy cones, not flowers, with seeds not borne inside a fruit.

Hair A threadlike outgrowth of the outermost layer of cells.

Hardy Tolerant of cold growing conditions.

Hermaphroditic Refers to a plant or flower that has both stamens and carpels in the same flower.

Hybrid A plant resulting from a cross between two species, subspecies, or cultivars. Hybrids occur naturally or can be deliberately bred. They are indicated with an × symbol in the scientific name.

Inflorescence A cluster of flowers along with any associated structures.

Introduced species A non-native that was purposely or inadvertently imported to grow in a certain region.

Invasive Refers to introduced plants that outcompete and displace native species.

Keel A centrally located ridge on the long axis of a structure, usually on the lower side, like the keel of a boat.

Leaf A usually photosynthetic organ arising from a stem, usually with a stalk and a blade.

Leaf arrangement The number of leaves attached at each node on the stem. Described as alternate (1 per node), opposite (2 per node), or whorled (3 or more per node).

Leaf form Refers to the blade of the leaf, e.g., simple, compound, or pinnate.

Leaflet A smaller, leaflike unit of a compound leaf.

Legume A flattish, beanlike pod, usually made by members of the bean or pea family (Fabaceae), generally splitting longitudinally along two lines.

Linear Long and narrow, with nearly parallel sides. Usually a linear leaf is six or more times longer than it is wide.

Lobe An expansion or bulge on the margin of a structure such as a leaf or petal. A division to about halfway of any organ.

Margin The edge of a leaf or other flat organ.

Mediterranean climate A climate characterized by cool, wet winters and warm, dry summers.

Membranous Thin, pliable, and often more or less translucent.

Midvein The promient vein in the center of a leaf.

Monoecious Refers to plants with unisexual male and female flowers that are separate but on the same plant.

Native An indigenous plant: one that occurs naturally in an area.

Naturalized Said of a non-native, often weedy plant that reproduces on its own without the help of humans.

Needle-like leaf A stiff, narrowly linear, usually evergreen leaf.

Node The position on a stem to which structures, often leaves, are or were attached.

Nut A dry, indehiscent fruit with a hard shell that usually encases a single seed.

Opposite Leaves that arise from a stem in pairs, two per node.

Ovary The ovule-bearing portion of a pistil, which develops into a fruit.

Ovule An immature seed. In gymnosperms exposed to the air at the time of pollination, in angiosperms enclosed within an ovary.

Palm A plant in the palm family (Arecaceae), having an unbranched trunk crowned by large pinnate or palmate leaves.

Palmate leaf A leaf whose veins, lobes, or leaflets radiate from a single point, like fingers from the palm of a hand.

Pedicel The stalk of an individual flower or fruit.

Peduncle Stalk of an inflorescence.

Pendent Hanging downward loosely.

Petal A usually colorful individual member of the corolla.

Petiole The stalk of the leaf, connecting the leaf blade to the stem.

Phyllode A leaf, such as those found in acacias, consisting of a flattened, bladelike petiole.

Pinnate leaf A leaf whose veins, lobes, or leaflets radiate on opposite sides of a central axis in a featherlike fashion.

Pistil The female reproductive structure of a flower, composed of an ovary tipped by one or more styles and one or more stigmas.

Pollen Grainlike structures containing male sexual cells, produced in the anther.

Pollination The transfer of pollen from male reproductive structures to female reproductive structures, leading to fertilization.

Pome Fleshy fruit in the rose family (Rosaceae), such as an apple or pear, in which the bulk of the flesh comes from enlarged receptacle or outer floral parts.

Prickles Sharp-pointed projections derived from the outermost layers of cells of a stem or leaf.

Propagate To generate new individual plants by seeds, grafting, or other methods.

Pulvinus The swollen base of a petiole that functions to permit leaf movement.

Rare A plant that occurs only in a small number of areas or as a few individuals.

Receptacle The enlarged base of a flower or cone stalk, to which reproductive parts are attached.

Root The underground structure of a plant that functions in anchorage, absorption and transport of water and nutrients, and food storage.

Samara A winged, one-seed fruit such as those produced by ash trees.

Scale-like leaf A small leaf, not differentiated into blade and petiole, often pressed tightly to the stem.

Seed A reproductive structure in angiosperms and gymnosperms consisting of an outer protective layer enclosing a dormant embryo and nutritive tissue; a ripened ovule.

Sepal An individual member of the calyx, generally green.

Serrate leaf A leaf with sawtoothed margins.

Sessile Refers to a leaf, flower, or fruit that lacks a stalk and is attached directly to the stem.

Shoot The year's new growth at the end of a branch.

Shrub A relatively short woody plant, generally with multiple trunks.

Simple leaf A leaf with a single, undivided blade.

Species A group of populations with interbreeding individuals that are related genetically and are often reproductively isolated from other such groups; the taxonomic category that ranks below genus.

Spike An unbranched inflorescence, often upright, with sessile flowers, which usually open from bottom to top of the spike.

Spine A sharp-pointed projection derived from a leaf.

Stamen The male reproductive structure of a flower, composed of a stalklike filament and a terminal, pollen-producing anther.

Stem The aboveground (occasionally belowground) axis of a plant, to which leaves and buds are attached.

Sterile Not bearing sex organs or not involved in reproduction.

Stigma On a pistil, the terminal, often sticky portion on which pollen is deposited.

Stipules A pair of leaflike structures (occasionally spine- or gland-like) at either side of the base of the petiole.

Street tree A tree planted along streets and sidewalks of a municipality, usually in parkways between sidewalks and curbs.

Style The stalklike section of a pistil that connects the ovary to one or more stigmas.

Subspecies A geographically and often physically distinct population within a species (abbreviated "subsp.").

Temperates Regions of Earth between the tropics and the polar circles.

Terminal At the end or tip of a structure.

Thorn A sharp-pointed projection derived from a stem or shoot.

Tooth On a leaf, a small, pointed projection.

Tree A tall, woody plant, usually with one large trunk.

Tropics Regions of Earth spanning the equator (below 23.4° north or south).

Truncate Terminating abruptly, as if cut straight across or nearly so.

Unisexual flowers Flowers with stamens or carpels, but not both.

Urban forest The sum of all the trees and associated living organisms in an urban or suburban area.

Variety A group within a species that differs in consistent but minor ways from the rest of the species, similar to subspecies.

Veins The strands of transport tissue often seen as lines in leaves.

Venation The arrangement of veins in a leaf, often described as parallel, pinnate, or palmate.

Weed An uncultivated, unwanted, nuisance plant growing in a human-made setting (e.g., agricultural field, park, or garden) or invading into undisturbed areas.

Weeping Drooping or bending downward.

Whorl A group of three or more leaves, floral parts, or flowers that are arranged in a ring around a central point or axis.

Wood The inner tissue of a tree, running from the roots to the branchlets, consisting of hard, mostly dead cells that conduct water and provide stiffness and structural support.

Woody A portion of a plant, usually a shrub or tree, that is hard and thickened, often covered in bark.

References, Further Reading, and Great Tree Books

Trees of Santa Barbara, by Robert N. Muller and John Robert Haller (Santa Barbara, CA: Santa Barbara Botanic Garden, 2005).

Trees of Seattle, by Arthur Lee Jacobson (Seattle: Arthur Lee Jacobson, 2006).

The Trees of San Francisco, by Michael Sullivan (Petaluma, CA: Pomegranate, 2004).

The Trees of Golden Gate Park and San Francisco, by Elizabeth McClintock and Richard G. Turner (Berkeley: Heyday, 2001).

An Illustrated Manual of Pacific Coast Trees, by Howard McMinn, Evelyn Maino, and H. W. Shepherd (Berkeley: Univ. of California Press, 1935).

Manual of Cultivated Plants Most Commonly Grown in the Continental United States and Canada, by L. H. Bailey (New York: Macmillan, 1949).

Sunset Western Garden Book, by Kathleen Norris Brenzel (Birmingham, AL: Oxmoor House, 2007).

Horticultural Flora of South Eastern Australia, by Roger Spencer and Su Pearson (Sydney: Univ. of New South Wales Press, 2002).

The Tree, by Colin Tudge (New York: Random House, 2007).

Mabberley's Plant-Book, by D. J. Mabberley (New York: Cambridge Univ. Press, 1997).

The Urban Tree Book, by Arthur Plotnik (New York: Random House, 2000).

Firefly Encyclopedia of Trees, by Steve Cafferty (Buffalo: Firefly, 2005).

The Trees of San Diego, by Steve Brigham and Don Walker (San Diego: San Diego Horticultural Society, 2005).

Trees of Stanford and Environs, by Ronald N. Bracewell (Stanford, CA: Stanford Historical Society, 2005).

Index

Acacia spp. 9, 15, 19, 46–51, 57. *A. baileyana* 49, 50; *A. cultriformis* 46, 49; *A. cyclops* 46, 49; *A. dealbata* 46, 47, 49, 118; *A. decurrens* 48, 49; *A. longifolia* 46, 47, 48, 49; *A. mearnsii* 48, 49; *A. melanoxylon* 46, 47, 49, 51; *A. pendula* 49; *A. pycnantha* 49; *A. redolens* 49; *A. retinodes* 46, 49; *A. saligna* 48, 49; *A. stenophylla* 48, 49; *A. verticillata* 48, 49, 67; key 49

acacia: Bailey 49, 50; blackwood 46, 47, 49, 51; everblooming 46, 49; knife 46, 49; shoestring 48, 49; star 48, 49; weeping 49; key 49

acai 164

Acer spp. 17; *A. buergerianum* 53; *A. campestre* 52, 53; *A. macrophyllum* 53; *A. negundo* 14, 15, 53; *A. oblongum* 53; *A. palmatum* 52, 53, 157; *A. paxii* 53; *A. platanoides* 52, 53; *A. pseudoplatanus* 53; *A. rubrum* 53; *A. saccharinum* 52, 53; *A. saccharum* 52, 53; key 53

acorns 11, 105, 138–146

aerial roots 90, 92, 121

Aesculus spp. 7; *A. × carnea* 8, 9, 14, 54; *A. californica* 15, 35, 54; *A. hippocastanum* 14, 54; *A. pavia* 54

afghan pine 39, 41

African fern pine 14, 19, 28, 173

African sumac 14, 15, 149

African tulip tree 15, 60, 150

Afrocarpus falcatus 14, 19, 28, 173; *A. gracilior, see A. falcatus*

Agathis spp. 29

Agonis flexuosa 19, 55

agricultural trees 125, 136

Ailanthus altissima 16, 46, 56, 118, 167

Albizia julibrissin 9, 15, 57

alder: 24, 58; Italian 58; red 58; white 35, 58

Aleppo pine 39, 41

Alexandra palm 166

Allocasuarina verticillata 67

Alnus spp. 24; *A. cordata* 58; *A. rhombifolia* 35, 58; *A. rubra* 58

Aloe bainesii 13; *A. barberiae* 13

alternate leaves (illustrated) 7

Altingiaceae 110

American elm 160

American linden 155

American sycamore 130, 131

Anacardiaceae 127, 148, 148

Anacardium occidentale 127

Andean wax palm 164

angiosperms 45–162; defined 7

Angophora spp. 78; *A. costata* 18

anther, defined 9

apple, argyle 87, 88, 89

apple trees 116

apricot, Japanese 134

Araucaria spp. 13, 14, 29; *A. araucana* 19, 30; *A. bidwillii* 19, 29, 30, 51, 111, 167; *A. columnaris* 8, 13, 14, 30; *A. cunninghamii* 30; *A. heterophylla* 30; key 30

Araucariaceae 29–30

Arbutus 'Marina' 6, 9, 23, 59; *A. andrachne* 59; *A. canariensis* 59; *A. menziesii* 112; *A. unedo* 23, 59

Archontophoenix alexandrae 166; *A. cunninghamiana* 165, 166

Arecaceae 166–171

argyle apple 87, 88, 89

aromatic trees. *See* fragrant trees

ash: 7; flowering 95; green 95; Modesto 94, 95; Oregon 95; raywood 94, 95; shamel 94, 95; velvet, *see* ash, Modesto; white 95; key 95

athel tree 13, 154

Australian fan palm 165

Australian willow 19, 96

avocado 21, 69, 108

axillary bud, defined 6

baccalaureate, defined 108

Bailey acacia 49, 50

bald cypress 14

banana shrub 122

bank catclaw 49

banksia 98

banyan 90

bark 6

Bauhin, Johann and Gaspard 60

Bauhinia spp. 9, 12, 60; *B. × blakeana* 60; *B. forficata* 60; *B. variegata* 60

bay laurel 31, 108, 161

bay, sweet 21

bead tree 9, 6, 23

beech, European 23

beech family. *See* Fagaceae

Betula spp. 24, 58; *B. nigra* 6; *B. pendula* 61

Betulaceae 58, 61

Beverly Hills 92

Bidwill's coral tree 83

big leaf maple 53

Bignoniaceae 68, 74, 99, 102, 150

bipinnate leaf (illustrated) 7

birch: 24, 58; white 61

birch family. *See* Betulaceae

black locust 17, 118, 147, 167; hybrids 9

black mulberry 123
black oak 140
black wattle 48, 49
blackwood acacia 46, 47, 49, 51
blade, defined 7
blireiana plum 134, 135
bloom color and time (chart) 9
blue atlas cedar 32
blue fan palm 165
blue gum 46, 67, 84, 85, 87, 88, 89, 105, 111, 114, 118, 167
blue oak 139, 140
Bombyx mori 123
bottle tree: 20, 62–63; Queensland 20, 62, 63; key 63
bottlebrush: 18; lemon 64, 65; weeping 9, 64, 65
box: Brisbane 21, 112, 114, 173; Victorian 31, 46, 101, 128
box elder 14, 15, 53
Brachychiton acerifolius 9, 62, 63; *B. discolor* 9, 62, 63; *B. populneus* 8, 20, 62, 63; *B. rupestris* 20, 62, 63; key 63
bract, defined 155
Bradford, Frederick 136
Bradford pear. *See* Callery pear
Brahea armata 165; *B. edulis* 165
Brazilian coral tree 83
Brazilian pepper tree 17, 118, 148, 150
Brisbane box 21, 112, 114, 173
bristlecone pine 8
broad leaved coral tree 83
bronze loquat 23, 81
brush cherry 18, 153
buckeye 7, 8, 54; California 15, 35, 54
bud, defined 7
bud scales (illustrated) 6
bunya bunya 13, 14, 19, 29, 30, 51, 111, 167
bur oak 140, 141
Butia capitata 165
butterfly habitat (monarch) 84
cabbage palm, New Zealand 13
Caesalpinioideae 57
cajeput tree 118, 119, 150
Calamus spp. 164
California bay laurel 31, 108, 161
California buckeye 15, 35, 54
California fan palm 35, 164, 165, 170
California Invasive Plant Council 124
California lilac 22
California pepper tree. *See*

Peruvian pepper tree
California sycamore 35, 130, 132
Callery pear 9, 24, 96, 114, 136, 157
Callistemon spp. 18, 64–65, 121; *C. citrinus* 64, 65; *C. viminalis* 9, 64, 65
Callitropsis macrocarpa. See Hesperocyparis macrocarpa; C. nootkatensis 36
Calocedrus decurrens 13, 31, 32, 35
Calodendrum capense 18
calyx, defined 8
camphor tree 20, 31, 75, 105, 108, 111
Canary Island date palm 164, 165, 167, 168
Canary Island pine 39, 40
Cannabaceae 70
Cannabis sativa 70
canopy seed storage 35
canyon live oak 140, 141
cape chestnut 18
carat weight 71
carob tree 16, 71
Carolina laurel 21, 135
carpel, defined 9
Carpinus betulus 24
Carriere hawthorn 79
carrotwood 16, 17, 80, 118
Carya illinoinensis 16
Caryota spp. 165
cashew family. *See* Anacardiaceae
cashew trees 127
Cassia 57; *C. leptophylla* 9, 16, 66
Casuarina spp. 9, 13, 14, 105; *C. cunninghamiana* 67; *C. equisetifolia* 67; *C. stricta, see Allocasuarina verticillata; C. verticillata, see Allocasuarina verticillata*
Casuarinaceae 67
Catalina ironwood. *See* island ironwood
catalpa family. *See* Bignoniaceae
catalpa: southern 68; western 68, 74
Catalpa spp. 9, 17, 99; *C. bignonioides* 68, 74; *C. speciosa* 68; key 68
catalposide 68
catclaw, bank 49
cauliflory, defined 72
Ceanothus spp. 22
cedar: 14; atlas 32; blue atlas 32; Cyprus 32; deodar 8, 32, 111; incense 13, 31, 32, 35; of Lebanon 32; Port Orford

13, 32; red 32; true 32; western red 32; yellow 36
Cedrela sinensis 16
Cedrus spp. 14; *C. atlantica* 32; *C. atlantica* var. *glauca* 32; *C. brevifolia* 32; *C. deodara* 8, 32, 111; *C. libani* 32
Ceiba pentandra 69; *C. speciosa* 6, 9, 14, 60, 69, 105, 173
Celastraceae 117
Celtis spp. 23; *C. australis* 70; *C. laevigata* 70; *C. occidentalis* 70; *C. sinensis* 21, 70
Ceratonia 57; *C. siliqua* 16, 71
Cercidiphyllum japonicum 17
Cercis spp. 12, 57; *C. canadensis* 9, 20, 72; *C. occidentalis* 72; *C. siliquastrum* 72
Ceroxylon spp. 164
Chamaecyparis lawsoniana 13, 32
Chamaerops humilis 164, 165
cheese wood 31, 46, 101, 128
cherry: brush 18, 153; Cornelian 76; flowering 23, 60, 135; hollyleaf 21, 22, 23, 135; Japanese flowering 5, 9, 135; Kwanzan 135; Taiwan 135
cherry laurel 21, 135
chestnut, cape 18
Chilean wine palm 164, 165
Chilopsis linearis 18, 74
chinaberry 9, 16, 120
Chinese cedrela 16
Chinese elm 24, 114, 159, 160
Chinese flame tree 16, 104–105
Chinese fringe tree 18, 73
Chinese hackberry 70
Chinese juniper 37
Chinese parasol tree 22
Chinese photinia 23
Chinese pistache 17, 114, 127, 157
Chinese tallow tree 19, 21, 118, 157, 173
Chinese windmill palm 165, 169
Chionanthus retusus 18, 73; *C. virginicus* 73
Chiranthodendron pentadactylon 22
chitalpa 18, 74
×*Chitalpa tashkentensis* 18, 74
Chorisia speciosa. See Ceiba speciosa
Christmas tree, New Zealand 9, 18, 121
cider gum 88, 89
Cinnamomum camphora 20, 31, 75, 105, 108, 111; *C. zeylanicum* 75, 108
cinnamon 75, 108

citronella 77
citrus family. *See* Rutaceae
Citrus spp. 20, 22
clones 133
cloves 153
coast live oak 35, 46, 114, 138, 139, 140, 141, 142, 143
coast redwood 3, 8, 14, 35, 38, 44, 46, 111, 171
cockspur coral tree 82, 83
coconut 164
Cocos nucifera 164
common hackberry 70
common juniper 37
common names 173
compound leaf, defined 7
cone-forming trees 8
conifers, defined 7–8
Convention on International Trade in Endangered Species 120
Cook pine 8, 13, 14, 30
Cootamundra wattle 50
coral gum 84, 86, 88, 89
coral tree: 9, 57, 60, 82–83; Bidwill's 83; Brazilian 83; broad leaved 83; cockspur 82, 83; dwarf, *see* coral tree, natal; naked 82, 83; natal 83; pink 82, 83; South African 82, 83; Sykes 82, 83; Transvaal 83
Cordyline australis 13
cork oak 138, 139, 140, 145
Cornaceae 76
Cornelian cherry 76
Cornus spp. 9, 18; *C. capitata* 76; *C. florida* 76; *C. kousa* 76; *C. mas* 76; *C. nuttallii* 76
corolla, defined 8
Corylus spp. 24
Corymbia calophylla 78; *C. citriodora* 9, 19, 20, 31, 77, 88, 89, 173; *C. ficifolia* 9, 21, 60, 78, 86, 88, 89, 173
cow-itch tree 107
crab apple 9, 24, 31, 116
crape myrtle 9, 18, 21, 60, 106, 157
Crataegus spp. 24; *C.* × *lavallei* 22, 79; *C. laevigata* 22, 79; *C. phaenopyrum* 22, 79
creeping fig 91
Crimean linden 155
Crinodendron patagua 23
Cryptomeria japonica 13, 14
Cupaniopsis anacardioides 16, 17, 80, 118
Cupressaceae 31, 33, 35, 36, 37, 38, 44
Cupressus macrocarpa. See Hesperocyparis macrocarpa; C. sempervirens 13, 33

cypress: bald 14; Italian 13, 33; Leyland 13, 36; Mediterranean, *see Cupressus sempervirens*; Monterey 3, 13, 35, 36, 111, 167, 173
Cyprus cedar 32
cypress family. *See* Cupressaceae
date palm: 164, 165, 167; Canary Island 164, 165, 167, 168; pigmy 164, 165; Senegal 165
Davis 1
dawn redwood 14, 38
deciduous, defined 6
deodar cedar 8, 32, 111
dermatitis 127
Desert Museum palo verde 126
desert willow 18, 74
diamond leaf pittosporum 129
dioecious, defined 10
Diospyros spp. 3
DNA 173
Dodonaea viscosa 19
dogwood 9, 18; evergreen 76; flowering 76; Kousa 76; western 76
dogwood family. *See* Cornaceae
doubly serrate leaf margins in *Ulmus* 159
Douglas fir 14
Dracaena draco 13
dracaena, giant 13
dragon tree 13
drooping melaleuca 119
drought, adaptation to 47
Dutch elm 160; disease 159, 162
dwarf coral tree. *See* natal coral tree
eastern redbud 9, 20, 72
edible fig 21, 90, 91, 118, 167
edible loquat 23, 81
Elaeagnus angustifolia 118
Elaeis guineensis 164
elder, box 14, 15
elm: 23, 70, 111, 159–160, 162; American 160; Chinese 24, 114, 159, 160; Dutch 160; English 160; Scotch 160; Siberian 159, 160; key 160
empress tree 17
encina 142
Engelmann oak 139, 140, 141
English elm 160
English hawthorn 22, 79
English oak 140, 141
English walnut 17, 103
Equisetum spp. 67
Ericaceae 59
Eriobotrya deflexa 23, 81; *E. japonica* 23, 81
Erythrina spp. 9, 57, 60, 82–83;

E. × *bidwillii* 83; *E. caffra* 82, 83; *E. coralloides* 82, 83; *E. crista-galli* 82, 83; *E. falcata* 83; *E. humeana* 83; *E. latissima* 83; *E. lysistemon* 83; *E. speciosa* 82, 83; *E.* × *sykesii* 82, 83
Eucalyptus spp. 17, 20, 21, 46, 78, 84–89, 112, 121, 158; *E. camaldulensis* 46, 86, 87, 88, 89; *E. cinerea* 87, 88, 89; *E. citriodora, see Corymbia citriodora; E. cladocalyx* 88, 89; *E. conferruminata* 46, 88, 89; *E. diversicolor* 87; *E. erythrocorys* 87; *E. ficifolia, see Corymbia ficifolia; E. globulus* 46, 67, 84, 85, 87, 88, 89, 105, 111, 114, 118, 167; *E. grandis* 88; *E. gunnii* 88, 89; *E. leucoxylon* 6, 86, 87, 88, 89; *E. nicholii* 19, 84, 86, 88, 89; *E. polyanthemos* 85, 87, 88, 89; *E. pulverulenta* 86, 88, 89; *E. robusta* 86, 87, 88, 89; *E. rudis* 88, 89; *E. saligna* 88, 89; *E. sideroxylon* 6, 84, 86, 88, 89; *E. tereticornis* 88; *E. torquata* 84, 86, 88, 89; *E. viminalis* 88, 89; key 88. *See also Corymbia*
Euphorbiaceae 157
European beech 23
European hackberry 70
European hornbeam 24
European horse chestnut 14, 54
Euterpe oleracea 164
everblooming acacia 49
evergreen, defined 6
evergreen dogwood 76
evergreen maple 53
evergreen pear 9, 23, 24, 137
evolutionary relationships among trees 4, 78
Fabaceae 46–51, 60, 66, 71, 72, 82, 97, 126, 147, 152, 156; characteristics of 57
Faboideae 57
Fagaceae 138–146
Fagus sylvatica 23
fall color 110; tree list 157
fan palm: Australian 165; blue 165; California 35, 164, 165, 170; Mediterranean 164, 165; Mexican 114, 165, 170, 171
Feijoa sellowiana 17
fern pine 14, 19, 28, 173
Ficus spp. 19, 85, 91, 169; *F. auriculata* 90, 91; *F. benjamina* 91; *F. carica* 21, 90, 91, 118, 167; *F. elastica* 91;

F. macrophylla 5, 90, 91, 92, 111, 121; *F. microcarpa* 91, 93, 114; *F. pumila* 91; *F. rubiginosa* 90, 91; key 91

fig: 19, 85, 90–93, 167, 169; creeping 91; edible 21, 90, 91, 118, 167; Indian laurel 91, 93, 114; Moreton Bay 5, 90, 91, 92, 111, 121; rusty leaf 90, 91; weeping 91; key 91

fig family. *See* Moraceae

figwort family. *See* Scrophulariaceae

fir, Douglas 14

firewheel tree 9, 22, 151

Firmania simplex 22

fishtail palm 165

flame tree: 9, 62–63; Chinese 16, 104–105; Illawarra 9, 62, 63; key 63

flamegold tree 16, 104–105

flaxleaf paperbark 119

flooded gum 88, 89

floss silk tree 6, 9, 14, 60, 69, 105, 173

flowchart, tree identification 196

flower color and time (chart) 9

flower parts 8, 9–10

flowering ash 95

flowering cherry: 23, 60, 135; Japanese 5, 9, 135

flowering crab apple 9, 24, 31, 116

flowering dogwood 76

flowering plum 23

forest red gum 88

Formosan sweetgum 22

fossils 38

fragrant trees 101, 108; tree list 31

Franceschi, Francesco 104

frangipani 101

Fraxinus spp. 7; *F. americana* 95; *F. angustifolia* 'Raywood' 94, 95; *F. holotricha* 'Moraine' 95; *F. latifolia* 95; *F. ornus* 95; *F. pennsylvanica* 95; *F. uhdei* 94, 95; *F. velutina* 94, 95; key 95

fringe tree, Chinese 18, 73

fruit trees. *See* agricultural trees

fungal diseases 43

funiculus, defined 47, 51

Fusarium circinatum 43

galls, myoporum 124

Geijera parviflora 19, 96

giant dracaena 13

giant sequoia 8, 13, 38

giant yucca 13

Ginkgo biloba 12, 8, 34, 157, 169

Ginkgoaceae 34

Gleditsia 57; *G. triacanthos* 15, 16, 97

Glendora 92

glossy privet 18, 31, 46, 109

gold medallion tree 9, 16, 66

golden bough 143

golden trumpet tree 9, 14, 99

golden wattle 46, 47, 48, 49

golden wreath wattle 48, 49

goldenrain tree 9, 16, 104–105

Grecian laurel 21, 108

green ash 95

green wattle 48, 49

greenhouse gases xii

Grevillea robusta 9, 16, 98

growing conditions in California 2; urban 1

Guadalupe palm 165

guava, pineapple 17

gum: 17, 20, 21; blue 46, 67, 84, 85, 87, 88, 89, 105, 111, 114, 118, 167; cider 88, 89; coral 84, 86, 88, 89; flooded 88, 89; forest red 88; lemon scented 19, 20, 31, 77, 88, 89, 173; manna 88, 89; red 46, 86, 87, 88, 89; red cap 87; red flowering 9, 21, 60, 78, 86, 88, 89, 173; rose 88; silver dollar 85, 87, 88, 89; silver mountain 86, 88, 89; sour 21; spider 46, 88, 89; sugar 88, 89; Sydney blue 88, 89; water 19, 20, 21, 158, 173. *See also sweetgum*

gymnosperms 27–44

hackberry 23; Chinese 21, 70; common 70; European 70

Handroanthus spp. 60, 173; *H. chrysotrichus* 9, 14, 99; *H. heptaphyllus* 9, 15, 99

Harpephyllum caffrum 17

haws 79

hawthorn: 22, 24; Carriere 79; English 79; Lavalle 22; Washington 22, 79

hazel 24

heath family. *See* Ericaceae

heath melaleuca 6, 119

hedge maple 52, 53

height-to-diameter ratio of trees 171

Hesperocyparis macrocarpa 3, 13, 35, 36, 111, 167, 173

×*Hesperotropsis leylandii* 13, 36

Heteromeles arbutifolia 23, 100

hobo trees, list 46

holly 22, 143

holly oak 20, 138, 139, 140, 143

hollyleaf cherry 21, 22, 23, 135

Hollywood juniper 37

holm oak. *See* holly oak

homesteads of early California 167

honey locust 15, 16, 97

Hong Kong orchid tree 60

hoop pine 30

hop bush 19

hops 70

hopseed 19

hornbeam, European 24

horse chestnut: 7, 54; European 14; red 8, 9, 14

horsetail plant 67

horsetail tree 9, 13, 14, 67, 105

how this book works 10–11

Howeia spp. 165; *H. belmoreana* 164

Humulus lupulus 70

hybridization in *Corymbia* spp. 78

hybridization in magnolias 115

Hymenosporum flavum 9, 20, 31, 60, 101

identification flowchart 196

identification key, defined 11

Identification Key to Trees Commonly Cultivated in California 12–24

"Ilex," meaning of 143

Ilex spp. 22

Illawarra flame tree 9, 62, 63

incense cedar 13, 31, 32, 35

Indian laurel fig 91, 93, 114

inflorescence, defined 10

interior live oak 140

introduced species 2

invasive species 56, 80, 84, 98, 102, 109, 118, 120, 124, 128, 149, 148, 157, 166

ipê 99

iron bark: red 6, 84, 86, 88, 89; white 6, 86, 87, 88, 89

ironwood, Catalina. *See* island ironwood

ironwood, island 15, 113

island ironwood 15, 113

island oak 140, 141

Italian alder 58

Italian cypress 13, 33

Italian stone pine 5, 39, 42

jacaranda family. *See* Bignoniaceae

Jacaranda mimosifolia 6, 7, 9, 15, 60, 99, 102, 105, 150

Japanese apricot 134

Japanese black pine 39

Japanese cryptomeria 13, 14

Japanese flowering cherry 5, 9, 135

Japanese maple 52, 53, 157

Japanese pagoda tree 9, 16, 152, 173

Japanese red pine 39
Jelecote pine 39
Joshua tree 13
Jubaea chilensis 164, 165
Judas tree 72
Juglandaceae 103
Juglans spp. 16, 103; *J. californica* 103; *J. hindsii* 103; *J. regia* 17, 103
juniper: 13; Chinese 37; common 37; Hollywood 37
Juniperus spp. 13; *J. chinensis* 37; *J. communis* 37
kapok 69
karee 149
karo 20, 129
Karri 87
katsura tree 17
kentia palm 164, 165
key: defined 11; *Araucaria* 30; ashes 95; *Brachychiton* 63; catalpas 68; elms 160; eucalypts 88; *Ficus* 91; *Koelreuteria* 104; maples 53; *Melaleuca* 119; oaks 140; palms 165; pines 39; *Pittosporum* 129; *Prunus* 135; *Tilia* 155; trees commonly cultivated in California 12–24
king palm 165, 166
Klambothrips myopori 124
knife acacia 46, 49
Koelreuteria spp. 9, 104–105; *K. bipinnata* 16, 104, 105; *K. elegans* 16, 104; *K. paniculata* 9, 16, 104, 105; key 104
kohuhu 129
Kousa dogwood 76
kurrajong 8, 20, 62, 63
Kwanzan cherry 135
lacebark 9, 62, 63
Lagerstroemia spp. 9, 18, 21, 60, 106; *L. fauriei* 106; hybrids 157; *L. indica* 106
Lagunaria patersonia 9, 20, 107
largeleaf linden 155
largest urban trees, list 111
Latin names 173
Lauraceae 75, 108, 161
laurel: 108; California bay 31; Carolina 21, 135; cherry 21, 135; Grecian 21, 108; Portugal 22, 23, 135
laurel family. *See* Lauraceae
Laurus nobilis 21, 108, 161
Lavalle hawthorn 22
leaf arrangement (illustrated) 7
leaf forms (illustrated) 6, 7
leaf scar (illustrated) 6
leaflet, defined 7
legume, defined 46
legume family. *See* Fabaceae
lemon bottlebrush 64, 65

lemon scented gum 19, 20, 31, 77, 88, 89, 173
lemonade berry 149
lemonwood 129
lenticel (illustrated) 6
Leptospermum laevigatum 19, 46; *L. scoparium* 19
Leyland cypress 13, 36
Ligustrum japonicum 18; *L. lucidum* 18, 31, 46, 109
lilac: California 22; Persian 6, 9, 120; wild 22
lily magnolia 115
lily of the valley tree 23
linden: 22, 155; American 155; Crimean 155; largeleaf 155; littleleaf 155; silver 155; key 155
Linnaeus, Carl 60
Liquidambar formosana 22; *L. styraciflua* 22, 31, 105, 110, 114, 157
Liriodendron tulipifera 5, 9, 12, 111, 157
littleleaf linden 155
Livistona spp. 165
locust: black 9, 17, 118, 147, 167; honey 15, 16, 97
Lombardy poplar 22, 133
London plane tree 6, 114, 130, 131, 173
long-leaved yellowwood 28
Lophostemon confertus 21, 112, 114, 173
loquat: bronze 23, 81; edible 23, 81
Los Angeles 1, 2; Arboretum 66
Lyonothamnus floribundus subsp. *aspleniifolius* 15, 113
Lythraceae 106
Macadamia integrifolia 17, 98
madrone 112
Magnolia denudata 115; *M. grandiflora* 2, 8, 9, 20, 31, 105, 114, 122, 167; *M. liliiflora* 115; *M.* × *soulangeana* 9, 20, 60, 115
magnolia: lily 115; saucer 9, 20, 60, 115; southern 2, 8, 9, 20, 31, 105, 114, 122, 167; yulan 115
Magnoliaceae 111, 114, 115, 122
mahogany 120
mahogany family. *See* Meliaceae
mahogany, swamp 86, 87, 88, 89
maidenhair tree 8, 12, 34, 157, 169
mallee 85
mallow family. *See* Malvaceae
Malus spp. 24, 116; *M. communis*

nis 116; *M.* 'Donald Wyman' 116; *M.* × *domestica* 116; *M.* × *floribunda* 9, 24, 31, 116; *M.* 'Liset' 116; *M. pumila* 116
Malvaceae 62, 69, 107, 155
Mangifera indica 127
mango 127
manna gum 88, 89
maple: 17, 52–53; big leaf 53; evergreen 53; hedge 52, 53; Japanese 52, 53, 157; Norway 52, 53; red 53; silver 53; smooth leaf 53; sugar 52, 53; sycamore 53; trident 53; key 53
maple syrup 52
marijuana 70
Markhamia lutea 15
Marri 78
marula 127
massive trees 8
Mataco Indians 69
mayten tree 6, 117
Maytenus boaria 6, 117
medicinal plants 79, 84, 96, 104, 110, 117, 118
Mediterranean cypress. *See* *Cupressus sempervirens*
Mediterranean fan palm 164, 165
Melaleuca spp. 17, 18, 65, 121; *M. armillaris* 119; *M. ericifolia* 6, 119; *M. linariifolia* 119; *M. nesophila* 119; *M. quinquenervia* 118, 119, 150; *M. styphelioides* 119; key 119
Melia azedarach 9, 16, 120
Meliaceae 120
Metasequoia glyptostroboides 14, 38
Metrosideros excelsa 9, 18, 121
Mexican fan palm 114, 165, 170, 171
Mexican palo verde 13, 15
Mexican sycamore 130
Michelia doltsopa 20, 122; *M. excelsa*, see *M. doltsopa*; *M. figo* 33; *M.* × *foggii* 122
michelia, sweet 20, 122
Mimosoideae 57
missions: California 167; Mission San Diego 125; Mission San Luis Rey de Francia 148
mistletoe 95
mock orange 31, 46, 101, 128
Modesto ash 94, 95
monarch butterfly habitat 84
Mondell pine. *See* afghan pine
monkey hand tree 22
monkey puzzle 19, 30
monoecious, defined 10
Monterey cypress 3, 13, 35, 36, 111, 167, 173

Monterey pine 35, 39, 43, 111, 167
Moraceae 90–93, 123
Moreton Bay fig 5, 90, 91, 92, 111, 121
Morus alba 21, 123, 157; *M. nigra* 123
most widely cultivated urban trees, list 114
Mugo pine 39
mulberry: black 123; white 21, 123, 157
mulberry family. *See* Moraceae
Myoporum laetum 20, 22, 46, 118, 124
Myrtaceae 55, 64–65, 77, 78, 84, 112, 118, 121, 153; distinguishing characteristics 158
myrtle, crape 9, 18, 21, 60, 106, 157
myrtle family. *See* Myrtaceae
naked coral tree 82, 83
narrow-leaf peppermint 19, 84, 86, 88, 89
narrow-leaf pittosporum 19, 129
natal coral tree 83
National Arboretum, US 106
Native Californians 100, 144
native, how to define 169
native trees of California 3, 35, 52, 108, 112, 131, 138, 142, 144, 164, 170. *See also Acer, Aesculus californica, Alnus rhombifolia, Alnus rubra, Arbutus menziesii, Calocedrus decurrens, Cercis occidentalis, Cornus nuttallii, Fraxinus, Hesperocyparis macrocarpa, Juniperus* (including *J. communis*), *Pinus* (including *P. longaeva, P. radiata*), *Platanus racemosa, Quercus* (including *Q. agrifolia, Q. lobata*), *Sequoia sempervirens, Sequoiadendron gigantea, Washingtonia filifera*
needle-like leaves (illustrated) 6
Nerium oleander 17
New Zealand cabbage palm 13
New Zealand Christmas tree 9, 18, 121
ngaio 124
niaouli 118
nitrogen-fixing bacteria 147
node, defined 7
Nolina recurvata 13
Norfolk Island pine 14, 13, 30
North American red buckeye 54
Northern California black

walnut 103
Norway maple 52, 53
nut trees. *See* agricultural trees
Nyssa sylvatica 21
oak: 22, 23, 85, 105, 138–146; black 140; blue 139, 140; bur 140, 141; canyon live 140, 141; coast live 35, 46, 114, 138, 139, 140, 141, 142, 143; cork 138, 139, 140, 145; Engelmann 139, 140, 141; English 140, 141; holly 20, 138, 139, 140, 143; holm, *see* oak, holly; interior live 140; island 140, 141; Oregon 140, 141; pin 140, 141; red 1, 138, 140, 141; scarlet 140, 141, 157; Shreve 140; silk 9, 16, 98; southern live 138, 140, 141, 146; valley 25, 35, 139, 140, 141, 144; key 140
Oakland 142
oil dots in Myrtaceae 158
oldest trees 8
"old-timey" trees, list 167
Olea spp. 169; *O. europaea* 18, 125, 167
Oleaceae 73, 94, 109, 125
oleander 17
olive family. *See* Oleaceae
olive trees 18, 125, 167, 169
olive, Russian 118
opposite leaves (illustrated) 7
orange, mock 31, 46, 101, 128
orchid family 57
orchid tree 9, 12, 57, 60
Oregon ash 95
Oregon oak 140, 141
Oriental plane 130, 131
origins of California's urban trees (chart) 3
ovary, defined 9
ovule, defined 9
pagoda tree, Japanese 9, 16, 152, 173
painkiller 96
palm: 164–171; Alexandra 166; Andean wax 164; Australian fan 165; blue fan 165; California fan 35, 164, 165, 170; Canary Island date 164, 165, 167, 168; Chilean wine 164, 165; Chinese windmill 165, 169; date 164, 165, 167; fishtail 165; Guadalupe 165; kentia 164, 165; king 165, 166; Mediterranean fan 164, 165; Mexican fan 114, 165, 170, 171; pigmy date 164, 165; pindo 165; pony tail 13; queen 105, 165, 168; rattan 164; Senegal date

165; windmill 165, 169; key 165
palm family. *See* Arecaceae
palm oil 164
palmate leaf (illustrated) 7
Palo Alto 1
palo brea 126
palo verde: Desert Museum 126; Mexican 13, 15, 126
paperbark: 118–119; flaxleaf 119; prickly 119
Papilionoideae 57
parasol tree, Chinese 22
Parkinsonia spp. 60, 126; *P. aculeata* 13, 15, 126; *P. ×* 'Desert Museum' 126; *P. praecox* 126
Parrotia persica 24
Paulownia kawakamii 17; *P. tomentosa* 17
pear, Bradford. *See* Callery pear
pear: Callery 9, 24, 96, 114, 136, 157; evergreen 9, 23, 24, 137; fruiting 136
pecan 16
pepper tree: 167; Brazilian 17, 118, 148; California, *see* pepper tree, Peruvian; Peruvian 16, 46, 148; simple-leaved 19
pepper (true) 148
peppercorns, pink 148
peppermint: narrow-leaf 19, 84, 86, 88, 89; willow 19, 55
bisexual flower, defined 10
Persea americana 21, 108
Persian lilac 9, 6, 120
Persian parrotia 24
persimmon 3
Peruvian pepper tree 16, 46, 148, 167
petal, defined 8
petiole, defined 7, 170
petticoat palm 170
Peyri, Anonio 148
Phoenix canariensis 164, 165, 167, 168; *P. dactylifera* 164, 165, 167; *P. reclinata* 165; *P. roebelenii* 164, 165
photinia, Chinese 23
Photinia serrulata 23
phyllode, defined 47
phylogeny 4, 78, 173
pigmy date palm 164, 165
pin oak 140, 141
Pinaceae 32, 39–43
pindo palm 165
pine: 14, 39–43, 67; afghan 39, 41; Aleppo 39, 41; bristle-cone 8; Canary Island 39; Cook 8, 13, 14, 30; distinguishing characteristics 85; fern 14, 19; hoop 30; Italian stone 5, 39, 42; Japanese

black 39; Japanese red 39; Jelecote 39; Mondell, *see* pine, afghan; Monterey 35, 39, 43, 111, 167; Mugo 39; Norfolk Island 13, 14, 30; Ponderosa 39; Torrey 39; Wollemi 29; key 39
pine family. *See* Pinaceae
pine nuts 42
pine pitch canker 43
pineapple guava 17
pink coral tree 82, 83
pink melaleuca 119
pink trumpet tree 9, 15, 99
pinnate leaf (illustrated) 7
Pinus spp. 14, 67, 85; *P. canariensis* 39, 40; *P. densiflora* 39; *P. eldarica* 39, 41; *P. halepensis* 39, 41; *P. longaeva* 8; *P. mugo* 39; *P. patula* 39; *P. pinea* 5, 39, 42; *P. ponderosa* 39; *P. radiata* 35, 39, 43, 111, 167; *P. thunbergii* 39; *P. torreyana* 39; key 39
Piper nigrum 148
Pissard plum 134
pistache, Chinese 17, 114, 127, 157
pistachio 127
Pistacia chinensis 17, 114, 127, 157; *P. vera* 127
Pittosporaceae 101, 128
Pittosporum spp. 21; *P. angustifolium* 19, 129; *P. crassifolium* 20, 129; *P. eugenioides* 129; pittosporum, narrowleaf 19; *P. rhombifolium* 129; *P. tenuifolium* 129; *P. tobira* 129; *P. undulatum* 31, 46, 101, 128; key 129
plane tree 22; London 6, 114, 130, 131, 173; Oriental 130, 131
plantation timber 43, 84, 147
Platanaceae 130–132
Platanus spp. 22, 130; *P. × hispanica* 6, 114, 130, 131, 173; *P. mexicana* 130; *P. occidentalis* 130, 131; *P. orientalis* 130, 131; *P. racemosa* 35, 130, 132
plum: blireiana 134, 135; flowering 23; Pissard 134; purple leaf 9, 114, 134, 135; South African 17
Plumeria rubra 101
Podocarpaceae 28
Podocarpus gracilior, see Afrocarpus falcatus; *P. henkelii* 28; *P. macrophyllus* 19, 28
poison ivy 127
poison oak 127, 149
pollarding 131

pollination 68, 82, 90, 98, 101, 107, 114
pollution xii, 1
Ponderosa pine 39
pony tail palm 13
poplar, Lombardy 22, 133
Populus nigra 'Italica' 22, 133
Port Orford cedar 13, 32
Portugal laurel 22, 23, 135
Pretoria 102
prickly paperbark 119
pride of India 9, 6, 120
primrose tree 9, 20, 107
privet: 18; glossy 18, 31, 46, 109
Proteaceae 98, 151
Prunus spp. 23, 60; *P. × blireana* 134, 135; *P. campanulata* 135; *P. caroliniana* 21, 135; *P. cerasifera* 9, 114, 134, 135; *P. ilicifolia* 21, 22, 23, 135; *P. laurocerasus* 21, 135; *P. lusitanica* 22, 23, 135; *P. mume* 'Alphandii' 134; *P. serrulata* 9, 135; *P. serrulata* 'Kwanzan' 5, 135; *P. speciosa* 135; *P. speciosa* 'Kwanzan' 135; key 135
Pseudotsuga spp. 14
pulvinus, defined 57, 72
puriri 14, 15
purple leaf plum 9, 114, 134, 135
purple orchid tree 60
Pyrus calleryana 9, 24, 96, 114, 136, 157, *see also P. kawakamii*; *P. communis* 136; *P. kawakamii* 9, 23, 24, 137; *P. taiwanensis, see P. calleryana*
queen palm 105, 165, 168
Queensland bottle tree 20, 62, 63
Quercus spp. 22, 23, 85, 105, 138–146; *Q. agrifolia* 35, 46, 114, 138, 139, 140, 141, 142, 143; *Q. chrysolepis* 140, 141; *Q. coccinea* 140, 141, 157; *Q. douglasii* 139, 140; *Q. engelmannii* 139, 140, 141; *Q. garryana* 140, 141; *Q. hypoleucoides* 141; *Q. ilex* 20, 138, 139, 140, 143; *Q. kelloggii* 140; *Q. lobata* 25, 35, 139, 140, 141, 144; *Q. macrocarpa* 140, 141; *Q. palustris* 140, 141; *Q. parvula* var. *shrevei* 140; *Q. robur* 140, 141; *Q. rubra* 1, 138, 140, 141; *Q. suber* 138, 139, 140, 145; *Q. tomentella* 140, 141; *Q. virginiana* 138, 140, 141, 146; *Q. wislizeni* 140; key 140

railroad-track trees, list 46
rattan palm 164
Raven, Peter xi–xii
raywood ash 94, 95
red alder 58
red cap gum 87
red flowering gum 9, 21, 60, 78, 86, 88, 89, 173
red gum 46, 86, 87, 88, 89
red horse chestnut 8, 9, 14, 54
red iron bark 6, 84, 86, 88, 89
red maidenhair tree 157
red maple 53
red oak 1, 138, 140, 141
redbud 12, 57; eastern 9, 20, 72; western 72
red-eyed wattle 46, 49
redwood: coast 3, 8, 14, 35, 38, 44, 46, 111, 171; dawn 14, 38. *See also giant sequoia*
reproduction in trees 8–10
Rhus spp. 127; *R. lancea, see Searsia lancea*
roadway and railroad-track trees, list 46
Robin, Jean 147
Robinia spp. 9, 57; *R. × ambigua* 147; *R. pseudoacacia* 17, 118, 147, 167
Rosaceae 79, 81, 100, 113, 116, 134–135, 136, 137
rose family. *See* Rosaceae
rose gum 88
rubber tree 91
rue family. *See* Rutaceae
Russian olive 118
rusty leaf fig 90, 91
Rutaceae 96
Sacramento Valley 2
Salicaceae 31
Salix spp. 23, 96
samara, defined 52, 160
San Diego 2, 92
San Francisco 2, 44, 59, 92, 121, 158; San Francisco–Oakland Bay Bridge 2
San Joaquin Valley 2
Santa Barbara 1, 2, 92; Channel Islands 113; County Courthouse 3
Santa Cruz island ironwood. *See* island ironwood
Santa Monica 92
Sapindaceae 52–53, 54, 80, 104–105
Sapium sebiferum. See Triadica sebifera
Saratoga bay laurel 108
Saratoga Horticultural Foundation 59, 96
Sassafras albidum 108
sato zakura 135
saucer magnolia 9, 20, 60, 115

sawleaf zelkova 23, 24, 157, 162

scale-like leaves (illustrated) 6

scarlet oak 140, 141, 157

Schefflera actinophylla 14

Schinus molle 16, 46, 148, 167; *S. polygamus* 19; *S. terebinthifolius* 17, 118, 148, 150

Shreve oak 140

scientific names 78, 173

Sclerocarya birrea 127

Scotch elm 160

Scrophulariaceae 124

Seaforthia elegans. See Archontophoenix cunninghamiana

Searsia lancea 14, 15, 149

seed dispersal 37, 47, 52, 80, 107, 109, 117, 120, 128, 129, 156, 148

seed storage, canopy 35

Senegal date palm 165

sepal, defined 8

sequoia, giant 8, 13, 38

Sequoia sempervirens 3, 8, 14, 35, 38, 44, 46, 111, 171

Sequoiadendron giganteum 8, 13, 38

serrate leaf margins in *Ulmus* 159

shamel ash 94, 95

sheoak 9, 13, 14, 67, 105

sheoak family. *See* Casuarinaceae

shoestring acacia 48, 49

showiest trees, list 60

shrubs 3

Siberian elm 160

silk oak 9, 16, 98

silk tree 9, 15, 57

silkworms 123

silver dollar gum 85, 87, 88, 89

silver linden 155

silver maple 52, 53

silver mountain gum 86, 88, 89

silver wattle 46, 47, 49, 118

Simaroubaceae 56

simple leaf (illustrated) 7

simple-leaved pepper tree 19

Smith, Betty 56

smooth leaf maple 53

soapberry family. *See* Sapindaceae

Sophora japonica. See Styphnolobium japonicum

Soulange-Bodin, M. 115

sour gum 21

South African coral tree 82, 83

South African plum 17

Southern California black walnut 103

southern catalpa 68

southern live oak 138, 140, 141, 146

southern magnolia 2, 8, 9, 20, 31, 105, 114, 122, 167

Spathodea campanulata 15, 150

spider gum 46, 88, 89

spurge family. *See* Euphorbiaceae

stamen, defined 9

star acacia 48, 49

Stenocarpus sinuatus 9, 22, 151

stigma, defined 10

stipules, defined 90

strawberry tree 6, 9, 23, 59

style, defined 10

Styphnolobium japonicum 9, 16, 152, 173

sugar bush 149

sugar gum 88, 89

sugar maple 52, 53

sugarberry 70

sumac: 127; African 149

sumac family. *See* Anacardiaceae

sunflower family 57

swamp mahogany 86, 87, 88, 89

sweet bay 21, 108

sweet michelia 20, 122

sweetgum: 22, 31, 105, 110, 114, 157; Formosan 22

sweetgum family. *See* Altingiaceae

sweetshade 9, 20, 31, 60, 101

Swietenia 120

Syagrus romanzoffiana 105, 165, 168

sycamore: 22; American 130, 131; California 35, 130, 132; Mexican 130

sycamore maple 53

syconium, defined 90

Sydney blue gum 88, 89

Sykes coral tree 82, 83

Syzygium aromaticum 153; *S. australe* 18, 153; *S. paniculatum, see S. australe*

Tabebuia. See Handroanthus

Taiwan cherry 135

tallest flowering plant 85

tallest trees 8

tallow tree, Chinese 19, 21, 118, 157, 173

Tamaricaceae 154

Tamarix spp. 154; *T. aphylla* 13, 154

tannins 138

Taxodium spp. 14

taxonomy 4, 78, 173

Taxus spp. 14

tea tree 19, 46

terminal bud (illustrated) 6

terminal bud scale scar (illustrated) 6

Thousand Oaks 142

thrips, myoporum 124

Thuja plicata 32

Tilia spp. 22; *T. americana* 155; *T. cordata* 155; *T. × euchlora* 155; *T. platyphyllos* 155; *T. tomentosa* 155; key 155

timber harvesting xii, 43, 46, 52, 84, 120, 122, 147

tipu tree 9, 15, 17, 57, 156

Tipuana tipu 9, 15, 17, 57, 156

tobira 129

Torrey pine 39

Toxicodendron diversilobum 127, 149; *T. radicans* 127

toyon 23, 100

Trachycarpus fortunei 165, 169

Transvaal coral tree 83

tree aloe 13

tree of heaven 16, 46, 56, 118, 167

Triadica: distinguished from *Sapium* 157; *T. sebifera* 19, 21, 118, 157, 173

trident maple 53

trip, trees most likely to (list) 105

Tristania conferta, see Lophostemon confertus; T. laurina, see Tristaniopsis laurina

Tristaniopsis laurina 19, 20, 21, 158, 173

trumpet tree: 60, 99, 173; golden 9, 14, 99; pink 9, 15, 99; yellow 9

trunk prickles (on *Ceiba speciosa*) 69

tulip tree: 5, 9, 12, 111, 157; African 15, 150

tupelo 21

Ulmaceae 159–160, 162

Ulmus spp. 23, 111; *U. americana* 160; *U. glabra* 160; *U. × hollandica* 160; *U. minor* 160; *U. parvifolia* 23, 24, 114, 159, 160; *U. procera, see Ulmus minor; U. pumila* 159, 160; key 160

Umbellularia californica 31, 108, 161

umbrella tree 14

urushiol 127

USDA Plant Introduction Station 136

valley oak 25, 35, 139, 140, 141, 144

velvet ash. *See* Modesto ash

Ventura 92

Victorian box 31, 46, 101, 128

Vitex lucens 14, 15

walnut: 16, 103; English 17, 103; Northern California black 103; Southern California black 103

Washington hawthorn 22, 79
Washingtonia spp. 168; *W. filifera* 35, 164, 165, 170; *W. robusta* 114, 165, 170, 171
wasps 90
water gum 19, 20, 21, 158, 173
water shortages xii
wattle: black 48; Cootamundra 50; golden 46, 47, 48, 49; golden wreath 48, 49; green 48, 49; red-eyed 46, 49; silver 46, 47, 49, 118
wax palm, Andean 164
weediest trees, list 118
weeping acacia 49
weeping bottlebrush 9, 64, 65
weeping fig 91
western catalpa 68
western dogwood 76
western red cedar 32
western redbud 72
white alder 35, 58
white ash 95
white birch 61
white iron bark 6, 86, 87, 88, 89
white mulberry 21, 123, 157
white orchid tree 60
whorled leaves (illustrated) 7
wild lilac 22
wilga 96
willow: 23, 96; Australian 19, 96; desert 18
willow peppermint 19, 55
willow pittosporum 129
windmill palm 165, 169
wine corks 145
wine palm, Chilean 164, 165
Wollemi pine 29
Wollemia nobilis 29
Xylosma congestum 23
yellow cypress 36
yellow trumpet tree 9
yellowwood 28
Yerba Buena Island 2
yew 14
yew pine 19, 28
Yucca brevifolia 13; *Y. elephantipes* 13
yucca, giant 13
yulan magnolia 115
Zelkova serrata 23, 24, 157, 162

About the Author

Matt Ritter grew up in a small town in rural Mendocino County. He attended UC Santa Barbara to earn a bachelor's degree in microbiology and UC San Diego for a Ph.D. in plant developmental biology. He is a botany professor in the biology department at Cal Poly, San Luis Obispo, where he teaches courses in general biology, general botany, plant diversity, and ecology. He has authored a number of scientific papers about plants and contributed to botanical references including *The Jepson Manual: Higher Plants of California* and the Flora of North America project. He studies trees that escape cultivation, particularly eucalyptus. He is a woodworker, athlete, musician, gardener, and all-around likable guy.

Also by Matt Ritter:

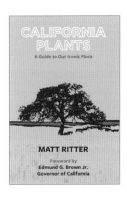
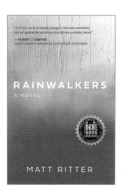

Tree Identification Flowchart

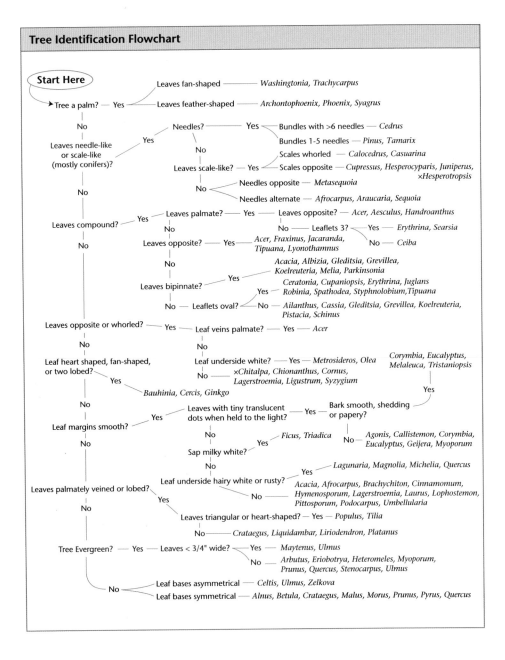